Studying Visual and Spatial Reasoning for Design Creativity

John S. Gero
Editor

Studying Visual and Spatial Reasoning for Design Creativity

Springer

Editor
John S.Gero
Krasnow Institute for Advanced Study
Fairfax
Virginia
USA

ISBN 978-94-017-7830-5 ISBN 978-94-017-9297-4 (eBook)
DOI 10.1007/978-94-017-9297-4

Springer Dordrecht Heidelberg New York London

Softcover reprint of the hardcover 1st edition 2015

Printed on acid-free paper

Springer is part of Springer Science+Business Media (www.springer.com)

Preface

Design has become a fruitful research area for cognitive scientists and computer scientists, not only for design scientists. For cognitive scientists it represents a particularly rich and open environment within which to study complex human behavior. For computer scientists it represents a challenging area. For design scientists the tools to study and model designers are only now becoming available. For neuroscientists design is largely a novel domain.

Design involves the creation of worlds and entails the interaction between minds and the representations of artefacts they produce. This opens a variety of areas for study since designers work with or without external tools, operate solely or within teams, and collaborate with others who are co-located or at a distance. Designers use external symbol systems extensively, particularly sketches, which implies a high engagement of visual and spatial reasoning.

Visual and spatial reasoning often play pivotal roles in design creativity, most noticeably through sketches, diagrams, visualization and visual imagery. There is research on visual and spatial reasoning in multiple disciplines but very little is focused on design creativity. It was to this gap that an NSF-funded workshop was organized in Provence, France to provide a forum to allow researchers in disparate disciplines to be exposed to the other's research methods and research results within the context of design creativity.

Representatives of the four disciplines of design science, computer science, cognitive science and cognitive neuroscience were invited to present and discuss their research and research methods. The workshop was structured so that the bulk of the time was spent on discussion. The four sessions leaders were:

- Design Science: Ramesh Krishnamurthi
- Computer Science: Christina Freksa
- Cognitive Science: Barbara Tversky
- Neuroscience: Jeff Zacks

The workshop provided a unique forum that brought together researchers from design science, computer science, cognitive science and cognitive neuroscience who were studying visual and spatial reasoning in their own ways, within an overarching framework of studying design creativity.

This volume contains fifteen of the papers presented and discussed at the workshop.

Pinelopi Kyriazi assisted in bringing the papers in this volume into a uniform whole, special thanks go to her. I would also like to thank Nathalie Jacobs, Anneke Pot and Cynthia Feenstra from Springer for all their assistance in bringing this volume to publication.

Krasnow Institute for Advanced Study John S. Gero

Contents

Contributors

Mehul Bhatt University of Bremen, Bremen, Germany

Evangelia G Chrysikou University of Kansas, Lawrence, KS, USA

James E. Corter Teachers College, Columbia University, New York, USA

R.C. Dalton University College London, London, UK

Max J. Egenhofer University of Maine, Orono, USA

Christian Freksa University of Bremen, Bremen, Germany

Gabriela Goldschmidt Israel Institute of Technology, Haifa, Israel

Center for Design Research, Stanford University, Stanford, USA

Ewa Grabska Jagiellonian University, Kraków, Poland

Tracy Hammond Texas A&M University, Texas, USA

C. Hölscher University of Freiburg, Freiburg, Germany

Maria Kozhevnikov Harvard Medical School, Charlestown, USA

National University of Singapore, Singapore, Singapore

Ramesh Krishnamurti Carnegie Mellon University, Pittsburgh, USA

Nora S. Newcombe Temple University, Philadelphia, USA

Jeffrey V. Nickerson Stevens Institute of Technology, Hoboken, USA

Juval Portugali Tel Aviv University, Tel Aviv, Israel

Amy L. Shelton The Johns Hopkins University, Baltimore, USA

Thomas F. Shipley Temple University, Philadelphia, USA

H.J. Spiers University College London, London, UK

Rudi Stouffs Delft University of Technology, Delft, The Netherlands

Barbara Tversky Teachers College, Columbia University, New York, USA

Stanford University, Stanford, USA

Jeffrey M. Zacks Washington University in St. Louis, St. Louis, USA

Part I
Design Science—State-of-the-Art

Navigating Complex Buildings: Cognition, Neuroscience and Architectural Design

R.C. Dalton, C. Hölscher and H.J. Spiers

This paper provides a tentative set of ideas which attempt to draw together research from neuroscience, spatial cognition and architecture (space syntax). It starts by considering the questions, "What does the brain do during the navigation of complex built space and how does it map it?" "What can cognitive studies tell us about navigation in complex buildings?" and "What does space syntax measure about structures of space and what does it tell us?" These questions serve as the starting point for the establishment of a framework for future collaborative efforts to bring together these disparate areas but with the fundamental aim of ultimately supporting architects to design more user-friendly buildings.

Introduction

In recent years, interest has been sparked by the overlap between architecture and neuroscience [1–4]. The majority of this research appears to focus on either one of two areas: how building users experience architecture and how this may affect them emotionally or, alternatively, on the creative process of architectural design: how architects design buildings and what happens at the neurological level, during the creative act. In Eberhard's 2009 article in Neuron [1], he describes the five broad areas that are studied in neuroscience as being: (1) sensation and perception, (2) learning and memory, (3) decision making, emotion and affect, and (4) movement or "*how do we interact with our environment and navigate through it?*" (Ibid. p. 755). It can be seen that the majority of research hitherto undertaken on the boundaries between neuroscience and architecture fit into the first and fourth of these areas.

R. C. Dalton (✉) · H. J. Spiers
University College London, London, UK
e-mail: ruth.dalton@northumbria.ac.uk

C. Hölscher
University of Freiburg, Freiburg, Germany

J. S. Gero (ed.), *Studying Visual and Spatial Reasoning for Design Creativity,*
DOI 10.1007/978-94-017-9297-4_1

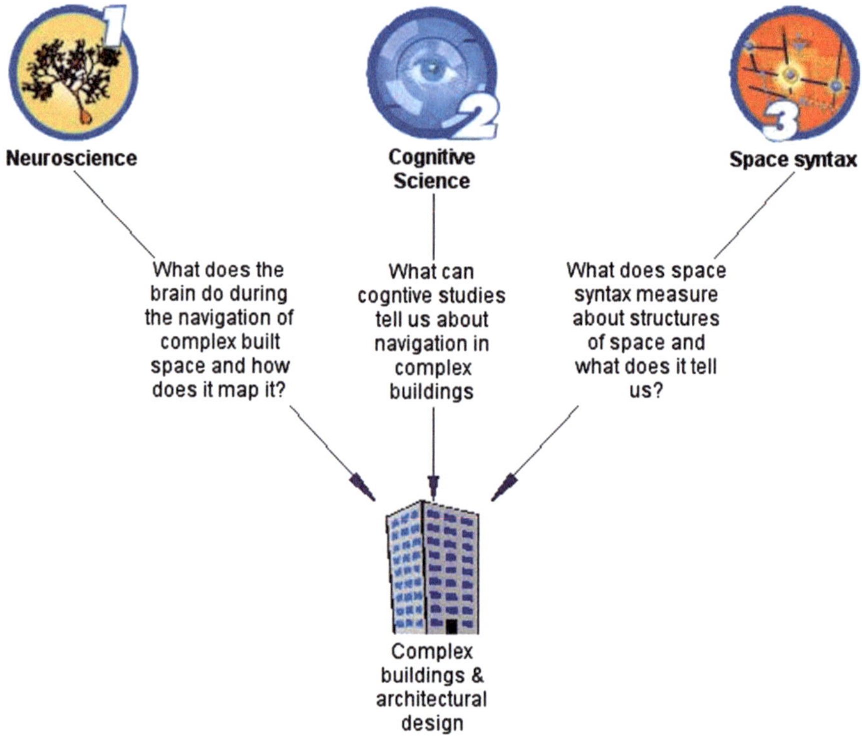

Fig. 1 The relative contributions of neuroscience, cognitive science and space syntax analysis to the design of complex buildings

This paper, in contrary to these other approaches, firmly concentrates on this latter area of study. To some extents, we are also addressing the third area, namely decision making, but only insofar as it is part of the act of wayfinding and navigation.

This paper will be presented in two sections; the first will examine three distinct and dissociated academic fields, neuroscience, cognitive science and space syntax (a discipline that emerged out of specifically architectural research, but arguably now covers a wider application domain), and will consider what recent developments in each of these fields can tell us about the usability and design of complex buildings, with a strong focus on issues such as wayfinding, navigability and legibility (see Fig. 1 for a diagrammatic representation of these three contributions). It will go on to suggest how these different strands of enquiry can be integrated and what potentials exist for future, collaborative research. The second section of this paper will examine the specific implications for the designer by suggesting ways in which these recent developments can be used to assist architects in the process of designing buildings that are more easily navigable and comprehensible, as informed from a spatial, cognitive and neuroscience standpoint. It is in this last section, that specific design issues will be addressed and we suggest how different approaches or heuristics might emerge from the empirical research in the neuroscience, cognition and architectural communities.

What Does the Brain Do During the Navigation of Complex Built Space and How Does It Map It?

If architects are called upon to design spatially complex environments that are effortlessly comprehensible, could it be of assistance to have an idea of what kind of neural activity takes place during navigation or even during the event of simply occupying a space? Until recently, such suppositions would have been purely speculative, but in recent years, through rapid advancements in neuroimaging, it is evident that a diverse network of brain regions are engaged during the navigation of complex built space. Navigation is a multi-faceted cognitive task, which relies on processing sensory information, coordinating movement, remembering the environment and planning. Thus, no wonder so many brain regions are active during navigation. Much has been learned from studies combining virtual reality environments and neuroimaging. Because neuroimaging requires the participant to remain very still inside the brain scanner, virtual reality has proved invaluable. Such studies have revealed that a network of brain regions including the hippocampus, parahippocampus, retrosplenial cortex, posterior parietal cortex and medial prefrontal cortex is more active during navigation [5–9]. Each of these regions is thought to serve a different purpose. Evidence suggests the hippocampus is responsible for storing an allocentric cognitive map of the environment to guide navigation. The parahippocampus may be important for processing topographical information necessary to determine current location and store information about the perceived environment. Retrosplenial cortex is thought to help translate the allocentric map information into egocentric information, which is processed in posterior parietal cortex to guide bodily movement or imagine movement during planning [8, 10]. The prefrontal cortex is important for planning routes, monitoring possible options and keeping the goal in mind during navigation [11, 12].

A detailed understanding the how the brain supports navigation and maps out the environment has come from studying the neurophysiology of the hippocampal formation and areas connected to it. Remarkably, cells in the hippocampal formation appear to contain an internal map and compass to support navigation. Evidence for this has come from recording the neuronal activity while an animal explores an environment. By continually recording the neuronal cell activity in the hippocampus along with the animal's position in an environment and it is possible to map the activity of cells to the surface of the environment and to the momentary orientation of the animal within it. This approach as revealed an elegant system dedicated to spatial mapping and orientation. Due to their distinctive properties cells in different regions of the hippocampal formation have been labeled with names such as 'place cells', 'head-direction cells', 'grid cells', and 'border cells'. The first to be discovered were place cells by O'Keefe and Dostrovsky in 1971 [13]. These exist in the hippocampus proper and fire action potentials (sending electrical signals to other cells) when an animal is in a particular location in the environment, but are typically silent otherwise. The location in an environment where a cell fires is called its place field: in a given environment only a subset of place cells will be active, with each cell's place field occupying a slightly different location, such that their collective, overlapping place fields carpet the whole environment. Place cells express different activity patterns in different environments, a phenomenon known as remapping. Place cells have several interesting properties. Their response appears to be a high-level conjunction of information that including knowledge of self motion. They respond predominately to changes in the boundaries, distant landmarks, and large-scale sensory aspects of the environment, such as the floor and wall colors [14]. They can learn over many trials to discriminate very similar environments [15].

Cells in a region next to the hippocampus called the presubiculum also produce a spatially tuned response, but it is not place related. Instead, these cells offer something akin to an internal compass by expressing activity tuned to certain head-directions in the current environment [16, 17]. Thus, one cell might fire maximally when an animal's head is facing northeast, another when facing southeast, another northwest, etc. Collectively the population covers all possible heading orientations. These cells are referred to as 'head-direction cells' and have also been found in other brain regions connected to the presubiculum, such as the anterior thalamus and retrosplenial cortex. The cells can be modulated both by vestibular and visual information. When prominent landmarks in an environment are rotated between visits to an environment these cells will tend to follow the rotation, with all cells rotating by the same amount. Thus these cells create a sort of internal compass, but not one which is oriented by magnetic fields.

'Grid cells' and 'Border cells' have both been discovered in the medial entorhinal cortex and subiculum. They are similar to place cells in that they show spatially localised patterns of activity in an environment, but they each differ from place cells in intriguing ways. Grid cell generate multiple place fields arranged in a tessellating grid-like pattern across the environment [18]. If lines are drawn between all fields, their pattern appears somewhat like a sheet of graph paper imposed on the environment, but rather than graph lines being at 90° right angles forming squares, the grid lines are at 60° to each other forming triangles. Simultaneously recorded grid cells show the same orientation of their grid pattern within an environment, but may show different spacing between fields. It is thought that grid cells provide inputs to place cells about the distance travelled in the environment. More recently it has been found that the horizontal structure of grid cells is not mimicked in their vertical structure (see the section on multi-level environments for more details), but rather they appear to be stacked in column-like structures [19].

Boundary vector cells [20] are also thought to provide inputs about the environment, and as their name suggests, they signal the location of borders in a given environment. Border cells will typically fire along, or just slightly offset to a border placed in an orientation matching its preferred orientation, e.g. Northwest. An important facet of the system is that in addition to the cells described, conjunctive cells exist, which combine grid or place properties with head-direction tuning [21]. These cells will only fire in a given place or set of places and only when an animal is facing in a particular direction. These have been found in the medial entorhinal cortex and presubiculum, but not the hippocampus proper.

Space is also mapped by cells in another part of the brain, the posterior parietal cortex [22]. Rather than mapping space relative to the environment, many of these cells code information relative to the body, forming egocentric spatial representations. Such cells respond to stimuli when placed at particular locations relative to, for example, the hand, torso, head, or eye [22]. These egocentric cells are organized to form structured maps of egocentric space, which can be used to remember where things are and guide actions through space to obtain goals, such as turning left or reaching/looking for an object. In order to create an internal map of an environment and navigate it, egocentric spatial information must be integrated with allocentric information, such that incoming sensory information acquired through egocentric receptors is integrated into a stable map, and in turn the map must be read out into egocentric referenced space to guide eye and limb movements.

What can Cognitive Studies Tell us About Navigation in Complex Buildings?

There are clear similarities and areas of overlap between neuroscience and cognitive science, not least of which is the primary focus of study, namely the human mind. Cognition refers to any of the 'higher-level' brain functions that begin to organize and structure the raw sense data, which represents our 'input' about our surroundings. Spatial cognition research, in particular, is concerned with the acquisition, organization, utilization, and constant revision of knowledge about spatial environments. One way for a layperson to understand what spatial cognition is about is that it is concerned with how "*that stuff out there*" (external to us), "*gets in here*" (is internalized in some manner, but at a far less physical, mechanistic level than would be of concern to the neuroscientist). As in any area of cognitive science, understanding both the underlying cognitive representation formats and the cognitive operations performed on such representations are key issues in spatial cognition, e.g., researchers on 'cognitive mapping' and wayfinding will be interested in both the representational formats of spatial information as well as in the mental operations that translate such information into navigation behavior or map drawing, but typically not concerned by the actual firing patterns of individual neurons.

Cognitive scientists have often created formal models of wayfinding behavior that allow larger structures and patterns to emerge (prominent examples include computational models [23, 24]. Other cognitive science researchers measure reaction time to investigate information processing. Environment and behavior researchers have developed tools such as sketch maps, think-aloud protocols, and the tracking of individuals to investigate aspects of human wayfinding such as identifying the different strategies used by people undertaking spatial navigation tasks and investigating the role of individual differences (particular in terms of spatial abilities) in wayfinding performance [25]. A recent taxonomy of human wayfinding tasks by [26] highlights the substantial differences between types of wayfinding tasks (e.g. exploring an environment vs. retracing a specific route), regarding both the information a person needs to have for successful task performance and their underlying spatial representation. Authors like Weisman [27] have provided taxonomic views of how perceptual features of the built environment (both indoor and outdoor) guide human wayfinding behavior and spatial learning. Weisman distinguishes four general classes of features: degree of visual access to the environment, architectural differentiation (i.e. visual similarity and thus ambiguity of locations along a route), complexity of a street or corridor layout as well as information conveyed by signs and maps. The space syntax approach introduced in the following section provides measures of such features, especially visual access and layout complexity.

So, what contributions have cognitive scientists made to research into the human navigation of complex environments, and what is currently known? First, there is a general consensus that there are three types of spatial knowledge: landmark knowledge, route knowledge and survey knowledge. Landmark knowledge is the identification and recollection of individual, distinct objects located in, and hence inextricably associated with a specific point in space. One way of viewing this is as a mental coupling between object and location (although the range of objects that can serve as landmarks is diverse, see below for further elaboration). In contrast, route knowledge concerns the storage and recollection of sequences of locations, each location being immediately and directly accessible from the previous location as well as to the subsequent one, such that they form a linked 'chain of associated locations'. Such sequences of adjacent places may be augmented by either directional and/or distance information. Finally, survey knowledge represents the most sophisticated form of spatial knowledge, since it concerns not only individual locations and their relative spatial interconnections, but must include additional information such as relative orientations of points in space and metric distances between places. It is this form of representation that is the most map-like of all the hypothesized forms of spatial knowledge. This progression, from least to most complex (landmark to route to survey knowledge) is also the order in which spatial knowledge is often thought to be acquired [28]. More recently authors like Montello [29] have challenged the notion of a strict progression from landmark to survey knowledge, pointing out that information on all three levels can be acquired in parallel and sometimes very rapidly. Nonetheless, coherent survey knowledge usually represents the ultimate state of greatest familiarity with an environment and hence tends to require the longest duration of attainment.

If we move from the more overarching framework of types of spatial knowledge presented above to research into the role of specific factors in people's understanding of their environment, then a number of distinct contributions to the field can be identified, which, when taken together form a coherent picture of how human spatial knowledge is attained, stored and subsequently used. For example, there is evidence that landmarks do appear play a role during navigation, although these are less generalizable or universally applicable (i.e. there is less of a 'one size fits all' explanation) than might have previously been theorized: salience of landmarks, individual differences and the particular form of wayfinding task all appear to play a role in people's selection, use and recall of landmarks [30, 31]. User's internal representations of environments, or cognitive maps, also tend to serve to simplify the external world and hence reduce its overall complexity. For example, slight deviations in routes or paths can frequently be recalled as being 'straighter' than they are in reality, with even some turns being omitted entirely. Routes with fewer turns are perceived of, or recalled, as being metrically shorter than those containing more changes in direction [32, 33]. Slight misalignments of spatial elements, for example rooms or corridors, are often canonicalised to cardinal angles. For example, an approximate 90° angle may be recalled as being a right-angle and approximate 'grids' can be regularized. In terms of multi-level environments (a large proportion of complex buildings will consist of more than one storey), Hölscher et al. [25] discovered an overriding assumption that subsequent floors will tend to resemble preceding floors, in terms of spatial layout and general spatial structure. When such assumptions are defeated, the resultant effect of wayfinding performance is measurable.

What Does Space Syntax Measure About Structures of Space and What Does it Tell us?

Space syntax is a set of theories, techniques and methods developed at UCL, London, in the 1970s which sought to describe the relationship between patterns of behavior and consequent, emergent social phenomena with objective, measurable properties of spatial systems. Unlike both neuroscience and cognitive science, space syntax research originated with questions about the nature of society rather than individuals, which, initially were of scant concern. Let us, therefore, start with society as a whole. One of the fundamental aspects of space syntax research has been the circular relationship between society and the types of spaces that they produce, namely that by studying the spaces produced by a society, we must surely be able to understand something about that society as a whole since spatial structures capture aspects of that society such as its values, power-relations, means of control and societal-hierarchies such as kinship structures. However, in turn, a society inhabits and is instantiated through those very same structures of space, which, in turn, have a direct effect upon all interpersonal interactions. In summary, society shapes space, which further shapes peoples' lives-as-lived.

With respect to this paper, however, perhaps we can begin to examine this relationship between society and its spaces of production a little differently. If we accept that humans, for the most part, create the spaces, which we inhabit, these spaces are, unequivocally, artifacts of human creation. Can it not be conjectured that if people conceptualize space in a certain manner, then these underlying spatial frameworks must somehow be encapsulated in the spatial systems we produce and inhabit, and, if so, they should equally be amenable to analysis? If a large enough sample of spatial systems (rooms, buildings, neighborhoods or cities) can be analyzed in such an objective manner that any underlying spatial commonalities can be clearly identified then such universalities might also be able to tell us something about how people conceptualize space. However, how do any such commonalities arise, given that the built environment is the product of so many minds and not simply the product of a single intelligence. This can be explained by Hillier's work on 'description retrieval'. He states, "... *It is proposed that there is [a general mechanism governing the link between geometric intuitions and spatial laws], and that it depends on the proposition that our mental interaction with the spatial world engages abstract relational ideas as well as concrete elements. In general, spatial relations are ideas with we think with rather than of.*" [34]. He uses the term 'description retrieval' to describe the emergent process of building cities in a piecemeal fashion, the process being one of first understanding or grasping the spatial rules of what is already there in order to add to and so reproduce those same spatial rules. It is a 'bottom-up' process that serves to maintain or even reinforce 'top-down' spatial rules. Given that such regularities or spatial commonalities can be observed, even across cultures, it is safe to assume that some process akin to Hillier's description retrieval must take place.

The other way of addressing the question posed in this section, "*What does space syntax measure about structures of space and what does it tell us?*" is to consider not merely the structures of space, but the ways in which people behave in those spaces, and whether the two are related. In order to understand the cyclical relationship between society and space, early space syntax researchers set out to observe and record aggregate patterns of spatial behavior such as occupancy and movement. It became rapidly evident that there was a strong and quantifiable relationship between aggregate flows of people through specific spaces and measures of how strategic that space was within the larger spatial system. In essence, the more 'integrated' the space (on average, how accessible a space is from all other spaces, measured mathematically) the more people are likely to pass through it. This strong relationship between space and movement (particularly pedestrian movement, but to a lesser extent, vehicular movement), has become a keystone of space syntax research, as its predictive power to estimate degrees of user flow- and occupancy-rates has become an invaluable tool for architects and urban designers wishing to evaluate schemes whilst still at a design stage. Naturally, such a wealth of observational data on aggregate pedestrian movement also has the potential to make a contribution to our understanding of how people cognize spatial systems.

Two additional areas of space syntax research that may contribute to research on human spatial cognition are 'angularity' and on 'intelligibility'. Angularity essentially represents a refinement of the basic space syntax analytic methods that considers not merely the topological relationship between two spaces (if two spaces are adjacent such that it is possible to pass unhindered from one to the other without passing through any intervening spaces, then they are considered connected in original space syntax analysis) but rather the physical angle turned through, or the change of direction taken, when passing from one space to another. This change in angle is represented as a set of 'weights' applied to the underlying graph-representation which underpins all space syntax measures. This refinement emerged from empirical work by Conroy Dalton, who observed that routes taken by subjects in a complex virtual environment appeared to favor more 'linear' and less meandering routes [35] and has subsequently been observed in GPS trails of London motorcycle

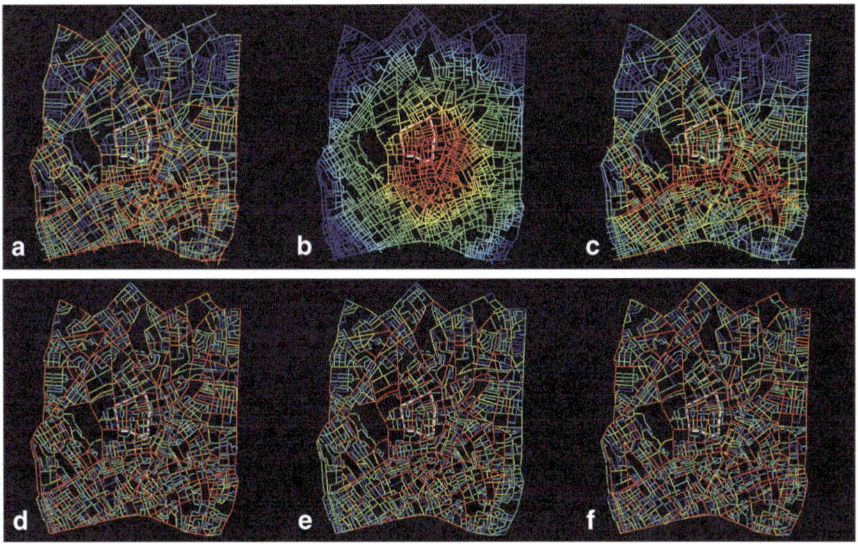

Fig. 2 Angular (**a,d**), metric (**b,e**) and topological (**c,f**) integration (*top row*) and choice (*bottom row*) values for the London neighborhood of Barnsbury (*red* = high values, *blue* = low value). (Source: Hillier and Iida [39])

couriers [36]. Early methodological work on how to re-conceptualize space syntax analytic techniques to include the concept of angular change originated with work by Dalton [37] and Turner [38]. Subsequent work by Hillier and Iida [39] served to quantify the increase in predictive power of the new angle-based analysis by using an observational dataset of pedestrian flow-rates taken from a wide variety of London neighborhoods and then correlating these to three measures of spatial structure: one using topological-based space syntax analysis, the second utilizing angle-weighted spatial graphs and finally a metric-based measure, in which the distance separating two points in space adds additional weights to the graph. When compared to the observational dataset, Hillier and Iida [39] found that the newer angle-based measures produced significantly higher correlations than the original topological measures or even the distance-based measures, which performed least well of the three. Figure 2 shows the plans for the neighborhood of Barnsbury and Fig. 3 shows the correlation coefficients for four districts and the shortest path (metric), least angle (angular) and fewest turns (topological) measures.

Intelligibility is another concept from Hillier [34] that suggests that our ability to find our way around a complex building or environment is partly dependent upon the relationship between local spatial variables and global spatial properties and hence our ability to draw inferences about one from the other. In an intelligible environment, suggests Hillier, spaces that are well connected will also tend to be highly strategic spaces within their larger spatial structure. These connections act as visual cues for the wayfinding pedestrian, since they can be easily discerned from the perspective of the situated observer. So, an intelligible environment, is one in

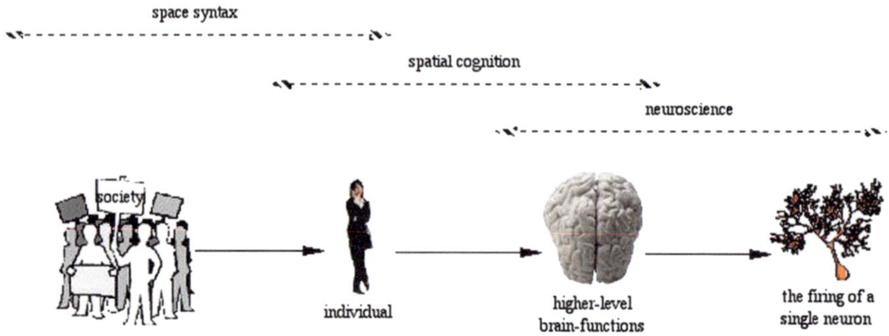

Fig. 3 The research-divide between society (as a collection of individuals), the individual, higher-level brain-functions & frameworks of knowledge, interactions between aggregations of neurons and different parts of the brain and the firing of a single neuron

Table 1 Correlation (r2) values for observed pedestrian movement counts against different path measures (shortest, least angle and fewest turns) for four districts in London. (Source: Hillier and Iida [39])

Neighbourhood	Pedestrian counts	Measure	Shortest path	Least angle	Fewest turns
Barnsbury	117	Accessibility Choice Combined	0.169 0.580 0.597	0.711 0.712 0.746	0.693 0.572 0.700
Calthorpe	63	Accessibility Choice Combined	0.114 0.440 0.534	0.586 0.552 0.615	0.605 0.364 0.605
South Kensington	87	Accessibility Choice Combined	0.116 0.342 0.383	0.561 0.480 0.589	0.551 0.540 0.615
Brompton	90	Accessibility Choice Combined	0.124 0.470 0.501	0.626 0.528 0.644	0.587 0.540 0.640
Mean Values across Neighbourhoods	–	Accessibility Choice Combined	0.143 0.458 0.471	0.621 0.568 0.649	0.609 0.504 0.640

which immediate, visual stimuli can provide cues about that which is beyond, and by definition outside, the immediate visual field. In contrast, in an unintelligible environment, the archetypal maze, for example, is one in which local visual cues do not relate to the larger spatial structure or, in the case of a maze, may be deliberately misleading. If this is true, do we actually utilize this spatial relationship to 'read' our environment and make inferences about the spatial structures around us? But, how do we internalize or make use of, consciously or unconsciously, this relationship between local and global patterns of space (the cognitive aspect) and, furthermore, what underpins this behavior at the neural level? (Table 1).

How Might These Strands be Integrated?

Having described some of the primary contributions of each of the three fields, neuroscience, spatial cognition and space syntax, to current work on pedestrian movement and navigation, this section will attempt to describe areas of synergy between the three approaches and where potentials for interesting interactions or future collaborations lie. First, however, we can immediately identify a problem created by the changes in the scale of focus, between these three academic fields. It can be seen from Fig. 3 that space syntax focuses on arguably the largest scale of all (society as a whole) and this change in scale continues until we reach the domain of neuroscience, which can focus on something as small as the firing of a single neuron.

So, how do we accommodate a shift from a preoccupation with society to the individual and then once more to the single neuron, and is the gulf that is required to be bridged between society and the individual greater or lesser than that between an individual's cognitive framework and their neural activity? One answer, to how to bridge such gulfs in scale, is to instigate collaborations on specific areas of research that initially appear to have the potential to make interesting contributions across two or more fields; areas where there already appear to be some connections or mutual relevance. We would like to initially suggest four areas, namely spatial knowledge acquisition, the role of orientation in wayfinding, multi-level environments (the third dimension) and navigation and intelligibility. The current connections are briefly described in the following sub-sections.

Spatial Knowledge Acquisition

Earlier in this paper, the three different types of spatial knowledge were described as being landmark knowledge, route knowledge and survey knowledge and it is further hypothesized, in psychology and cognitive science, that these representations can build upon each other and the spatial knowledge is often acquired in this order. It is suggested in this paper that there are interesting parallels between this sequence of knowledge acquisition and two other establish sequences: first the order in which space syntax measures correlate with observed pedestrian movement flows (angular measures correlating best, followed by topological distance and finally metric distance performing least well of all and second, work of Wills et al. [40] in the developmental stage of infant rats, on the relative maturation times of the different cells types, namely head direction cells maturing first, followed by place cells and finally grid cells. And, while it is not being suggested that a direct mapping can be drawn between, for example, research into the firing of head direction cells, angular distance and landmark knowledge (the top row in Fig. 4), it is being suggested that there might be some interesting research questions that emerge from placing these sequences in juxtaposition to one another. Is there a relationship between the how we acquire knowledge about environments and the order of maturation of cells involved in spatial orientation, for example, and between either of these sequences and observed aggregative movement rates in cities?

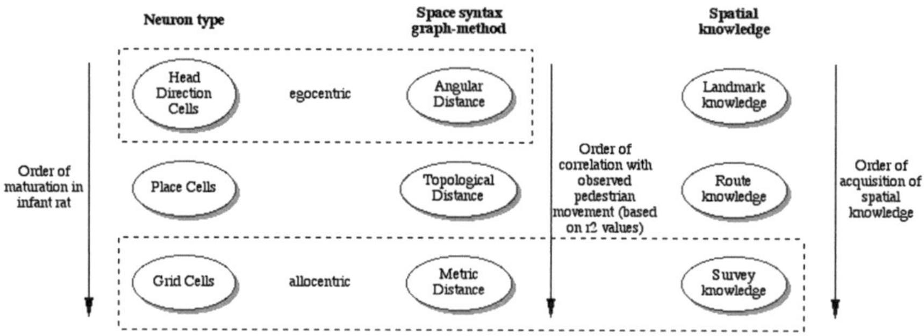

Fig. 4 *Left*, the order (*from top to bottom*) of maturation of cells in the infant rat; *middle*, the order (*from top to bottom*) of highest correlation with observed pedestrian movement; *right*, the order (*from top to bottom*) of acquisition of spatial knowledge. An indication is also given of which of these are fundamentally allocentric concepts and which are egocentric; those omitted have elements of both

Orientation

In the section on what the brain does during navigation, the function of head direction cells was described and it is worth remembering that they are not magnetic compasses since they do not respond the Earth's magnetic field. Like place cells and grid cells, they function in the dark and appear to be most closely linked to our vestibular system (in other words they are sensitive to head-turning) but there is evidence that our visual system serves to re-align our head direction cells regularly to compensate for natural 'drift'. In terms of investigating overlaps with areas of spatial cognition research, there are clear parallels with the 'route angularity effect' [32] and [33] in which routes that contain more or less changes of direction are judged to be shorter. It would be interesting to conduct experiments in carefully controlled virtual environments (containing routes of equal lengths and varying numbers/degrees of turns) in order to examine patterns of head cell firing in conjunction with any route angularity effects elicited. There are also clear areas of connection with Klippel et al.'s work on the canonicalisation of route directions, in particular with references to creating natural language expressions to describe directions or the schematization of map-representations of routes [41]. It would be fascinating to establish whether people's conceptions of a 'right turn' or the instructions to 'veer left' have any neurological basis in the firing patterns of head directions cells under in different environments. There are also clear areas of overlap with research from the space syntax community on angular-based measurements of spatial configuration and their strong predictive power on patterns of aggregate movement. One approach would be to modify the angular weightings currently used in space syntax analysis, such that the graph-weightings are far more aligned to human perceptions of angular change (ibid.) and then determine whether correlations with movement patterns improve. Another approach would be to extend Turner's work on using exosomatic visual agents [38], and attempt to give them not only simulated 'sight', as they currently possess, but also a similar sense of direction, provided by a set of simulated 'head direction cells.' Finally, an interesting possibility is to assess whether spatial syntax measures of angularity, topogological distance and metric distance have a correlate in the brain. Spiers and Maguire [42] scanned licensed London taxi drivers while they navigated a highly detailed simulation of London (UK). Activity in the entorhinal cortex and medial prefrontal cortex correlated with the metric (Euclidean) distance to the goal, measured at every second of the journey. Whether these, or other brain regions are also sensitive to the angularity and topological integration measures, remains to be investigated.

Multi-level Environments

What had remained an open question for some time after the discovery of grid cells was how they 'stacked up', i.e. what would be the effect of vertical movement? Did they also appear to form an equivalent, hexagonal, close-packing grid in the third dimension? This question has recently been solved by Verriotis et al. [19], when it was discovered that rat subjects, exploring a helical stair-like environment, produced radically different patterns of firing in the third dimension, i.e. it appears that the hippocampus encodes space differently in vertical and horizontal space. The pattern of grid cell firing appears to form a columnar-packing in the vertical dimension (see Fig. 6). One interpretation of this is, rather than our perception of space being three-dimensional, as such, it could rather be perceived of as being 2.5 dimensional, at best (assuming that grid cells in human brains are sufficiently similar to rat brains). This finding aligns particularly neatly with recent work by Hölscher et al. [25, 43] on the navigation of complex, multistory buildings, where he discovered that subjects tend to assume that different floors, stacked vertically, will more or less resemble each other, and when a building is encountered that radically departs from this model (i.e. Subsequent floors do not resemble lower floors in terms of general layout and arrangement, see Fig. 5 for an example of such an environment) then subjects can become rapidly disorientated. The finding that vertical space is encoded differently in the brain also serves to counter an occasional but reoccurring criticism of space syntax methods, namely that such methods are essentially two-dimensional and are therefore unable to sufficiently address the third dimension in buildings. The response has typically been that humans navigate in two dimensions rather than three and Verriotis et al.'s recent work on grid cells appear to substantiate this claim [19].

However, it is clear that additional work needs to be conducted into multilevel environments from the perspective of all three domains, neuroscience, spatial cognition and space syntax. In Montello's recent paper on the contribution of space syntax to environmental psychology, he raises a number of areas of future research, "In the future, space syntax will be expanded to include aspects of the third dimension in places, including the effects of vertically extended visual spaces on aesthetics and other responses, and the effects of vertical relationships in multi-level

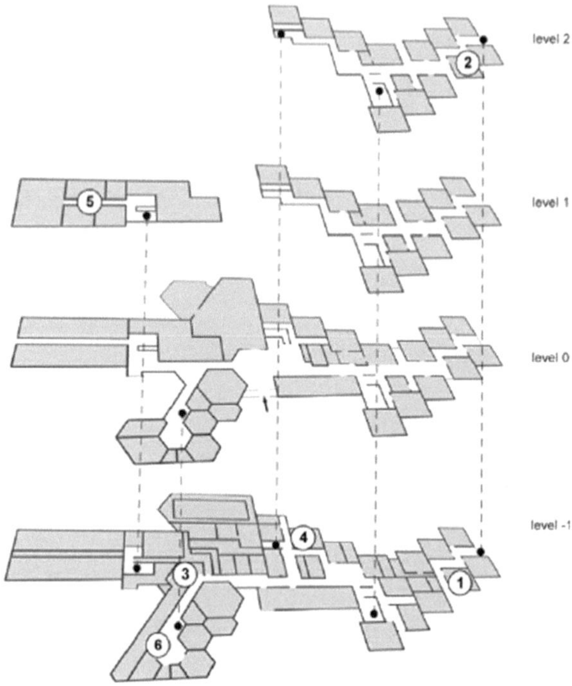

Fig. 5 Stacked, diagrammatic floorplan of the conference centre site used by Hölscher et al. for their study of multi-level wayfinding. Note the disparities between different floors [43]

structures on orientation and spatial learning." [44]. Tackling this problem in a unified way rather than as separate disciplines could best approach this manifesto of research into the effects of verticality.

Intelligibility

As mentioned in the section on space syntax, intelligibility is a concept from Hillier [34] that suggests that our ability to find our way around a complex building or environment is partly dependent upon the relationship between local spatial variables and global spatial properties and hence our ability to draw inferences about one from the other. Here lies a really interesting epistemological question about whether it is possible to infer global spatial properties from purely local spatial or visual ones and whether the found correlations (in space syntax analyses) between small-scale and large-scale spatial properties are either meaningful or can be actively utilized during navigation. One possible way of linking these different properties would be through a process of Hebbian synapse firing (in which any pair of cells that happen to fire simultaneously will evolve to strengthen their interconnections. In their paper on whether place cells can be connected by Hebbian synapses, Muller and Stead (1996) demonstrate how a simulation of a sequence of place cells along a route can form a synaptic chain (through a Hebbian process) and can produce a via

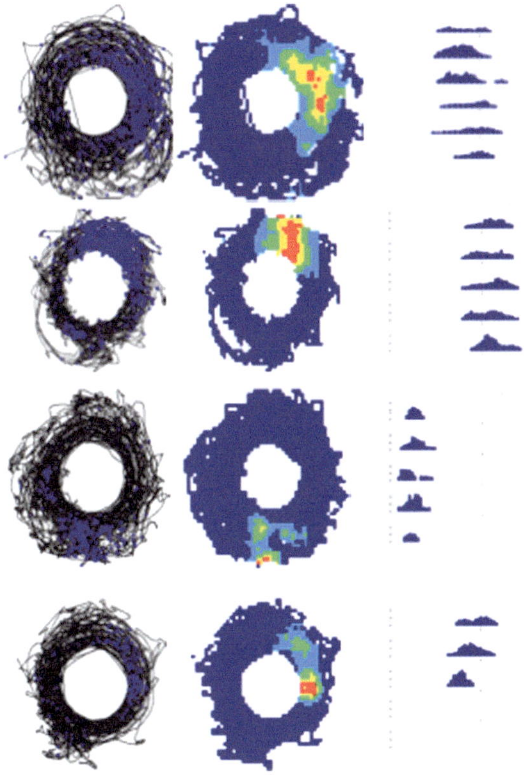

Fig. 6 *Left*: Four plan views of paths (*black line*) taken by a rat through a *helical* environment, *blue dots* indicate where a single gird cell was active in each of the paths. *Middle*: Plots of the firing rate for the cells shown left; *red regions* indicate where the maximum activity was located. *Right*: 4 sets of graphs of the activity of the each of the four cells. Each plot is from a different coil, arranged from *bottom* to *top* of the stairwell unwound and linearized, with the firing rate of the cell shown in *blue*. Note the activity of the cells is similar on all coils indicating the grid cells did not discriminate height of the rat in the environment. (Source: Hayman et al. [51])

route from an origin to a destination, hence making a connection between the local and immediate place and a distal goal location. The Muller and Stead model [45] is a very simple example of how this (inferences about a distal location from purely local information) might be achieved at a neurobiological level, therefore a similar process (but at a much larger scale of complexity) could begin to account for a process not unlike that of Hillier's 'intelligibility'. However, this would certainly be a learned response and therefore spatial cognitive research into spatial learning and infant development would be crucial to solving the question of how we make judgments about what we can not see from what we can, as would the contribution of space syntax to incorporate more accurate descriptions of spatial environments into any experimental method. Carlson et al. [46] have recently introduced an approach of combining highly controlled spatial learning experiments with naturalistic wayfinding studies to untangle contributing factors of spatial inference processes. Here inference is considered to be driven by higher-level cognitive heuristics and strategies rather than simple associative learning. This approach also builds upon the notion of intelligibility and space syntax measures are used as an integral part of designing test environments.

Implications for Architectural Design

In this second part of the paper we pose the question: what are the implications for the design of complex buildings, such as airports or hospitals that are regularly castigated as being disorientating and stressful environments? How could such information help an architect to design buildings that are intelligible and easily navigable? As Hillier states [34], architecture is both a rational and an intuitive/creative act and that each aspect is needed to result in a realizable building. If we imagine a future, where many of the questions posed above have been answered, even then would architects be any more able to design buildings that were easy for people to find their way around? The challenge is not merely one of knowing what information might be useful to the designer but also how best it might be made available (In what form and at what stage?). We suggest that the architect could be assisted by being provided with a set of analytic tools, guidelines and design-heuristics (emerging from research into wayfinding and navigation) to support the creative process. This evidence that is currently emerging from the disparate fields of neuroscience, cognitive science and space syntax research, and is already providing a basis for design: evidence-based design. However, this information coming from the different sources needs to be unified and the process accelerated into a clear program of research that aims not only to more fully understand the needs of users of architectural buildings but also to be able to help architects to put themselves into their end-users' shoes (known as 'perspective taking' in psychology and linguistics), by providing them with appropriate information.

We can begin to provide an idea of how such research could be translated into design-heuristics for architects to act as a checklist for designing for pedestrian movement. (Equally such rules-of-thumb could also be translated into fitness functions for generative designs.) However, the aim of any future program of research, integrating neuroscience, cognitive science and architecture, would be to expand and develop this list:

Design Heuristics for Architects

Straighter, more direct, routes are significantly preferable to routes containing many changes of direction. We recognize in this recommendation a potential conflict between aesthetics and wayfinding requirements: it can be tempting to 'break-up' a long corridor to create 'places' along the route. This may work aesthetically at a local level but will certainly hinder wayfinding [47].

Ensure unimpeded lines of sight connecting entrance spaces and other key, central spaces such as atria to the means of vertical circulation: stairs, lifts and escalators. These sight lines are crucial, so it is worth checking these explicitly on plan and/or using software (for example space syntax programs) designed to calculate such lines of sight [27, 34, 48].

Where changes of direction/orientation are unavoidable, shallower angles of turning (closer to straight-on) are preferable to sharp turns (and in particular try to avoid turning angles greater than 90°; forcing a building user to turn back can be disorienting).

Wherever possible ensure that differences in layout between floors are not too great. Building users will assume that each floor is laid out in an analogous manner to the preceding floors. Deviating from this too greatly will cause undue confusion [25].

When navigating outdoors, invariant visual information such as the horizon, position of sun, slope of ground or distal landmarks such as mountains can provide invaluable orientation checks whilst navigating. In a building, equivalent invariant information can be provided by ensuring frequent and regular sightlines to features such as external views, atria, or visually prominent architectural features.

Atria can serve another useful purpose: they can provide a 'short-cut' to survey-knowledge (or top-down and global as opposed to eye-level and local), as they facilitate views to and hence knowledge of other floors that would otherwise be unavailable. This bears some similarity to the concept of 'view enhancing' or the recommendation to climb a tree or another vantage point if lost outdoors [49, 50]. Atria can provide such 'view enhancement' opportunities to building users.

Excessive complexity should be avoided. Again, here lies potential for conflict between architectural intent and wayfinding functional requirements. Techniques are available, such as space syntax analyses, to check for overly complex designs.

Building users may become lost or disorientated in locations that bear strong visual similarity, at a local level, to other locations in the same building (this can often occur in strongly symmetrical layouts, i.e. one corridor is identical to a neighboring, parallel corridor) [27]. One design technique is to distinguish such locations through non-spatial means, such as prominent use of color. However, relying on internal décor as a navigational cue is problematic as this can unwittingly be altered over the life-time of the building; far better to avoid such spatially self-similar locations at the design-phase. Another method is to deliberately utilize architectural features as 'landmarks'. These work best if placed at decision points and if they are have a large visibility-catchment area, and so can be seen from multiple locations [30, 31].

Conclusions

This paper constitutes a tentative set of ideas that attempts to draw together research from neuroscience, spatial cognition and architecture (space syntax). This can hopefully serve as a springboard for future collaborative efforts in bringing together these areas but with the clear aim of supporting architects to design more user-friendly buildings.

Acknowledgements To Peg Rawes (Senior Lecturer, Bartlett School of Architecture, UCL) and Kate Jeffery (Professor and Director of Institute of Behavioural Neuroscience, UCL) for jointly organizing the "Spatial Thinking: visualising spatial thinking in architecture and neuroscience" one-day, interdisciplinary seminars in February 2010, which sparked off many of the ideas presented in this paper and to John O'Keefe (Professor of Cognitive Neuroscience, UCL) for some of his comments during the event, which were equally as inspirational.

References

1. Eberhard JP (2009a) Applying neuroscience to architecture. Neuron 62(6):753–756
2. Eberhard JP (2009b) Brain landscape: the coexistence of neuroscience and architecture. Oxford University Press, New York
3. Eberhard JP (2007) Architecture and the brain: a new knowledge base from neuroscience architecture. Oxford University Press, New York
4. Mallgrave HF (2010) Architect's brain: neuroscience, creativity and architecture. Wiley Blackwell
5. Maguire EA, Burgess N, Donnett JG, Frackowiak RS, Frith CD, O'Keefe J (1998) Knowing where and getting there: a human navigational network. Science 280(5365):921–924
6. Hartley T, Maguire EA, Spiers HJ, Burgess N (2003) The well-worn route and the path less travelled: distinct neural bases for route following and wayfinding in humans. Neuron 37(5):877–88
7. Spiers HJ, Maguire EA (2006) Thoughts, behaviour and brain dynamics during navigation in the real world. Neuroimage 31(4):1826–1840
8. Spiers HJ, Maguire EA (2007) Neuroscience of remote spatial memory: a tale of two cites. Neuroscience 149(1):7–27
9. Rauchs G, Orban P, Balteau E, Schmidt C, Degueldre C, Luxen A, Maquet P, Peigneux P (2008) Partially segregated neural networks for spatial and contextual memory in virtual navigation. Hippocampus 18(5):503–518
10. Byrne P, Becker S, Burgess N (2007) Remembering the past and imagining the future: a neural model of spatial memory and imagery. Psychol Rev 114(2):340–375
11. Ciaramelli E (2008) The role of ventromedial prefrontal cortex in navigation: a case study of impaired wayfinding and rehabilitation. Neuropsychologia 46(7):2099–2105
12. Spiers HJ (2008) Keeping the goal in mind: prefrontal contributions to spatial navigation. Neuropsychologia 46(7):2106–2108
13. O'Keefe J, Dostrovsky J (1971) The hippocampus as a spatial map. Preliminary evidence from unit activity in the freely-moving rat. Brain Research 34(1):171–175
14. O'Keefe (2007) The hippocampus book. Oxford University Press, Oxford
15. Lever C, Wills TJ, Caccuci F, Burgess N, O'Keefe J (2002) Long-term plasticity in hippocampal place-cell representation of environmental geometry. Nature 416(6876):90–94
16. Taube JS, Muller RU, Ranck JB Jr (1990a) Head-direction cells recorded from the postsubiculum in freely moving rats. I. Description and quantitative analysis. J Neurosci 10(2):420–435
17. Taube JS, Muller RU, Ranck JB Jr (1990b) Head-direction cells recorded from the postsubiculum in freely moving rats. II. Effects of environmental manipulations. J Neurosci 10(2):436–447
18. Hafting T, Fyhn M, Molden S, Moser MB, Moser EI (2005) Microstructure of a spatial map in the entorhinal cortex. Nature 436(7052):801
19. Verriotis M, Hayman R, Jovalekic A, Fenton AA, Jeffery KJ (2010) Anisotropic encoding of three-dimensional space by place cells and grid cells. Soc. Neurosci Abstr
20. Lever Burton S, Jeewajee A, O'Keefe J, Burgess N (2009) Boundary vector cells in the subiculum of the hippocampal formation. J Neurosci 29(31):9771–9777

21. Sargolini F, Fyhn M, Hafting T, McNaughton BL, Witter MP, Moser MB, Moser EI (2006) Conjunctive representation of position, direction, and velocity in entorhinal cortex. Science 312(5774):680–681
22. Andersen RA, Buneo CA (2002) Intentional maps in posterior parietal cortex. Annu Rev Neurosci 25:189–220
23. Chown E, Kaplan S, Kortenkamp D (1995) Prototypes, location and associative networks (PLAN): towards a unified theory of cognitive mapping. Cogn Sci 19:1–52
24. Kuipers B (1978) Modeling spatial knowledge. Cogn Sci 2:129–153
25. Hölscher C, Brösamle M, Vrachliotis G (in press) Challenges in multilevel wayfinding: a case study with the space syntax technique. Environment and Planning B: Planning and Design, 36
26. Wiener JM, Büchner SJ, Hölscher C (2009) Taxonomy of human wayfinding tasks: a knowledge-based approach. Spat Cogn Comput 9:152–165
27. Weisman J (1981) Evaluating architectural legibility: way-finding in the built environment. Environ Behav 13:189–204
28. Siegel AW, White SH (1975) The development of spatial representations of large-scale environments. In: Reese HW (ed) Advances in child development and behavior. vol 10. Academic, New York
29. Montello DR (1998) A new framework for understanding the acquisition of spatial knowledge in large-scale environments. In: Egenhofer MJ, Golledge RG (eds) Spatial and temporal reasoning in geographic information systems. Oxford University Press, New York, pp. 143–154
30. Nothegger C, Winter S, Raubal M (2004) Selection of salient features for route directions. Spat Cogn Comput 4(2):113–136
31. Presson CC, Montello DR (1988) Points of reference in spatial cognition: stalking the elusive landmark. Br J Dev Psychol 6:378–381
32. Sadalla EK, Magel SG (1980) The perception of traversed distance. Environ Behav 12(1):65–79
33. Sadalla EK, Montello DR (1989) Remembering changes in direction. Environ Behav 21:346–363
34. Hillier B (1996) Space is the machine: a configural theory of architecture. Cambridge University Press, Cambridge
35. Conroy Dalton R (2003) The secret is to follow your nose: route path selection and angularity. Environ Behav 35(1):107–131
36. Turner A (2009) The role of angularity in route choice: an analysis of motorcycle courier GPS traces. In: Stewart Hornsby K, Claramunt C, Denis M, Ligozat G (eds) Spatial information theory. Lecture notes in computer science: theoretical computer science and general issues (5756). Springer, Germany, pp. 489–504
37. Conroy Dalton R (2001) Spatial navigation in immersive virtual environments. Doctoral thesis, University of London
38. Turner A, Penn A (2002) Encoding natural movement as an agent-based system: an investigation into human pedestrian behaviour in the built environment. Environ Plan B: Plan Design 29(4):473–490
39. Hillier B, Iida S (2005) Network and psychological effects in urban movement. In: Cohn AG, Mark DM (eds) Proceedings of spatial information theory: international conference, COSIT 2005, Ellicottsville, N.Y., U.S.A., Sept 14–18, 2005. Lecture notes in computer science. vol 3693. Springer, Berlin, pp. 475–490
40. Wills TJ, Caccuci F, Burgess N, O'Keefe J (2010) Development of the hippocampal map in preweanling rats. Science 328(5985):1573–1576
41. Klippel A, Dewey C, Knauff M, Richter K, Montello DR, Freksa C, Loeliger E (2004) Direction Concepts in Wayfinding Assistance Systems. In: Jörg B, Christian K, Robert P (eds) Workshop on artificial intelligence in mobile systems (AIMS'04), SFB 378 Memo 84. Saarbrücken, pp. 1–8

42. Spiers HJ, Maguire EA (2007b) A navigational guidance system in the human brain. Hippocampus 17(8):618–26
43. Hölscher C, Meilinger T, Vrachliotis G, Brösamle M, Knauff M (2006) Up the down staircase: wayfinding strategies in multi-level buildings. J Environ Psychol 26:284–299
44. Montello D (2007) The Contribution of space syntax to a comprehensive theory of environmental psychology. Proceedings of 7th International Space Syntax Symposium, Istanbul
45. Muller RU, Stead M (1996) Hippocampal place cells connected by Hebbian synapses can solve spatial problems. Hippocampus 6:709–719
46. Carlson L, Hölscher C, Shipley T, Dalton RC (in press) Getting lost in buildings. Current Directions in Psychological Science
47. Conroy Dalton R (2005) Space syntax and spatial cognition. In: Wu D (ed) World architecture: Space Syntax.
48. Gärling T, Böök A, Lindberg E (1986) Spatial orientation and wayfinding in the designed environment: a conceptual analysis and some suggestions for postoccupancy evaluation. J Archit Plan Resour 3:55–64
49. Hill KA (1998) Lost person behavior. National SAR secretariat. Canada
50. Berdik C (2009) Why humans can't navigate out of a paper bag. New Scientist issue 2721
51. Hayman R, Verriotis M, Jovalekic A, Fenton A, Jeffery K (in press) Differential encoding of vertical and horizontal space by place cells and grid cells

Showing Connection

Jeffrey V. Nickerson, Barbara Tversky and James E. Corter

Introduction

Information systems exist in a space based on connection: Weightless data stream through wires and float through air. Some parts of the system are visible—processors, routers, disks. But the things that flow move so fast they can't be seen.

In system design, there is no site plan, no elevation, no perspective drawing, and so one might expect little visualization and spatial reasoning to supplement symbolic language. But while information systems designers speak words and write code, they also make use of many of the techniques that originate in other design domains [1, 2], first and foremost diagramming [3–6]. What do and what should information systems designers visualize? In order to address this question in manageable parts and discover areas for future research, we will describe the practice of design as an interaction with a set of increasingly abstract spaces.

First, the *geographic space* that we inhabit. Designers can locate parts of a system using GPS coordinates. But only a little is learned about a system from the geographic locations of the components. Instead, systems are commonly described in *network space*, which shows the structural and temporal connections between components. Stepping up a level of abstraction, the design process itself is a network, each design the product of a series of linked decisions. Designs are imagined as points in *design space,* the space of all possible designs. While design space is a place in designers generate alternatives, these alternatives need to be judged in *evaluation space*. The dimensions in this space are design objectives. Creativity involves a shuttling between points in the design space and evaluation space: alternatives are compared against objectives, and new designs are generated by making new decisions that fill out desirable regions of the evaluation space. The designs themselves are often represented as networks, which are mapped into representations on the

J. V. Nickerson (✉)
Stevens Institute of Technology, Hoboken, USA
e-mail: jnickerson@stevens.edu

B. Tversky · J. E. Corter
Teachers College, Columbia University, New York, USA

J. S. Gero (ed.), *Studying Visual and Spatial Reasoning for Design Creativity,*
DOI 10.1007/978-94-017-9297-4_2

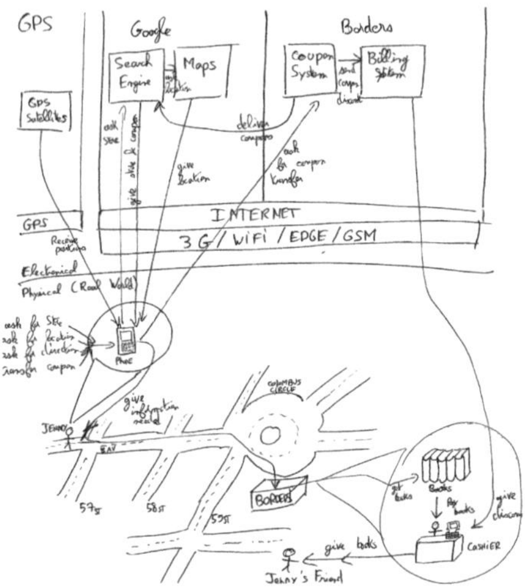

Fig. 1 A novice designer sketch for a system to distribute electronic coupons for a local bookstore. The geographic information at the bottom of the diagram is linked to the topological information at the top of the diagram

page, a part of geographic space. Hence the designer reasons in a variety of different spaces. The movement between spaces in the design process is seldom predetermined, but instead a result of situation, contingency, expertise, and style.

In the rest of the paper, we will discuss each space in more detail. We will point out places where research on visualization and spatial reasoning might further our understanding of design activity.

Geographic Space

When information systems are represented in geographic space, it is usually in order to show the location of particular components of a system: the network routers, the wires, the computers. Such locations are unimportant in many design situations. That is, even after installation, computers and network components can usually be moved without affecting the functions of the system. This movement affects the speed-of-light communication so little that differences in speed are undetectable.

There are, however, some situations in which geography is important. In the design of computer circuit boards, physical distance between wires can affect timing, heat dissipation, and the amount of radio interference. At a higher level, plans for wireless networks often show access points, and indicate with a circle the radio range within which other computers can connect.

Representation of systems in geographic space is usually straightforward, as the components can be placed on maps. An example of a design sketch that incorporates geography is shown in Fig. 1: the designer is showing how a bookstore might provide prospective customers electronic discount coupons. This example illustrates a broader phenomenon: wireless communication is giving renewed importance to

Table 1 Different forms of networks used in designing information systems. The last column, S/D, indicates if the network is meant to represent structure or dynamics

Nodes	Link basis	Representation	S/D
Computers	Wired or wireless network connections	Network diagram	S
People	Frequency of communication	Social network	S
Documents, people	Approval process	Workflow diagram	D
Software components	Calling structure	Call trees	D
Software objects	Inheritance	Class diagram	S
States of objects	Events	State diagram	D
Actors	Messages	Sequence diagram	D
Software, hardware	Software installed on hardware	Deployment diagram	S

location. The visualization conventions for wireless networks are still evolving, and present opportunities for research on both the production and understanding of diagrams addressing mobility.

Network Space

In contrast to geographic space, network space has no coordinates. Instead, a thing is described solely in terms of its connections. These very abstract structures can be used to describe many different phenomena, and with respect to information systems, there are a wide range of network descriptions, both standard ones taught to practitioners [7] and informal ones invented and reinvented to fit particular purposes [8]. These descriptions can be roughly classified into those that describe structure—what connects to what—and those that describe dynamics—the connecting order. Table 1 lists some of the many types of networks used to describe information systems.

Structural networks have as nodes people, computers, and software applications. These nodes have identifiers, and can be located in relation to neighbors, but there are seldom fixed locations. Figure 2 shows a typical sketch: the components are connected by lines that indicate interaction.

Other networks are constructed to indicate dynamics. In such networks, links often indicate the flow of a particular message from place to place. For example sequence diagrams, shown in Fig. 3a, show messages and the order of messages flowing from actor to actor in a system.

The vertical dimension of the diagram in Fig. 3a indicates time, whereas the horizontal dimension shows an instantaneous event, a message being passed. From traces of such messages, it is possible to specify the interfaces of an actor in the system. That is, just as one can specify many of the duties of managers in companies by watching their interactions, the interactions of software components the messages provide guidance on how they should be built. Software tools can help automate this process, and it is normal for design tools to facilitate this. But current tools are

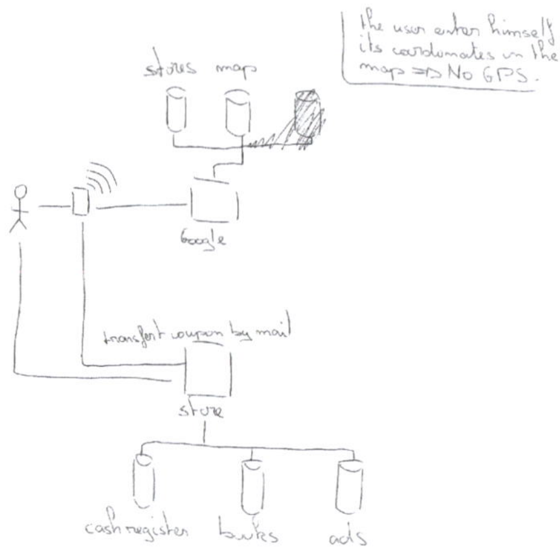

Fig. 2 Another example of the bookstore coupon problem, drawn by a different designer as a network

less useful in the next phase, when designers decide how to structure the connections between components. To what extent can the optimal structure of a network can be automatically inferred from the intended interactions shown in a diagram? One exploratory approach uses techniques from the psychological similarity literature [9]: Fig. 3b shows such an inference [10].

Time can also be combined in a single diagram, as shown in Fig. 4. These diagrams are effective at making inferences about how long a sequence of overlapping tasks will take, as the positions of the nodes in the tree correspond to elapsed time [11]. Are such combined diagrams easier or harder to infer from than a set of diagrams, such as those shown in Fig. 3? On the one hand, multiple diagrams demand designers perform a difficult integration task in the mind, but on the other hand, they avoid confounding structural and temporal aspects of a problem. This issue is important because increasingly computer systems rely on multiple processors

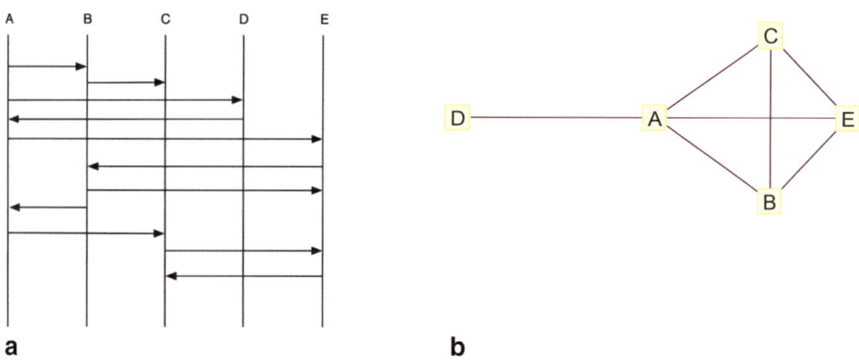

Fig. 3 On the left, **a** a sequence diagram: arrows indicate messages running between the actors of the system, *A*, *B*, *C*, *D* and *E*. On the right, **b** shows a network that is consistent with the sequence diagram, derived automatically using a technique discussed in [9]

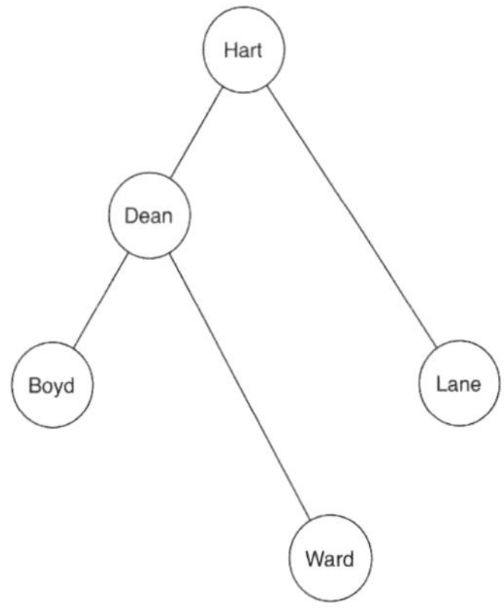

Fig. 4 A representation of a calling tree, in which the locations of the nodes indicate the elapsed time of the calls

running tasks in parallel. Designers need to decide which tasks can be run in such a way, and which must be run sequentially, and these decisions can have large effects on the performance and reliability of a system.

How do we reason about abstract networks? Network diagrams assume designers are capable of disregarding spatial information to focus on only on topological information, but this assumption may be optimistic. Studies of interactions with network diagrams have shown that distance along a line in the diagram is perceived more readily than topological distance [12]. Moreover, errors in enumerating paths on a network correlate with distance [13]. And even the positions of nodes in a network, information that should be arbitrary, are often uniform, based on cultural associations: for example, most designers will show, as in Fig. 1, a network provider above a store [6]. Consequently, we wonder the extent to which Euclidean bias leads to inferior decision making in the design and diagnosis of systems. It may be that training focused on topology will improve design and comprehension of systems, or it may be that we need different representations for systems that take into account our cognitive apparatus.

The tools created in order to facilitate systems design, Computer Aided Software Engineering tools, are not used very much in practice [14]. Programmers stick with a process of informal diagram sketching followed by coding in word processors. Are their tools that are less proscriptive, that can help a programmer without imposing onerous restrictions? Some think that tools that encourage reflective practice [15] will outdo existing tools by providing a gentler kind of guidance [16].

Even as software engineers struggle to find better ways of representing existing systems, emerging technologies present opportunities for new kinds of visualization that combine network and geographic space. For example, roads today can be mapped, using sensor data, to show the average speed on a particular portion of the road. Roads can also be mapped according to their connectivity to wireless access points or cell phone towers. It becomes possible to reason about how to traverse a road system while maintaining connection [17, 18].

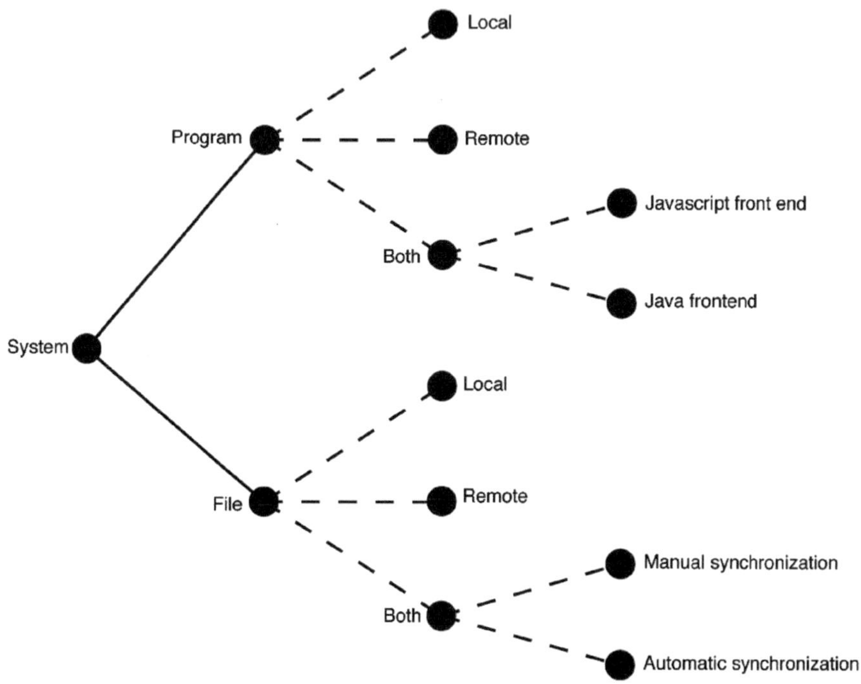

Fig. 5 A design tree for a system (for example a word processing application) that might have both local and cloud components. The tree is based on Brooks [3]: *Solid lines* indicate independent questions that all must be answered, and *dotted lines* indicate a mutually exclusive choice: only one path can be followed

Design Space

Design involves a progressive series of decisions [19]. Understanding the decision tree can be important for two reasons: it can aid in the generation of new designs, and it can help record design decisions, which can be useful for joining members of a design team.

How, then, should the decision tree be imagined? Brooks recommends what he calls a *design tree*, in which design questions are represented as nodes. Nodes send out either independent edges or mutually exclusive edges. All independent edges need to be followed. Mutually exclusive edges force a choice: only one of these several edges should be followed to create a design [3].

For example, imagine creating a design for a new word processor. The problem can be broken up into how to design the program, and how to design the storage of files. The program might reside locally, on the cloud, or both. Likewise, the data may reside locally, on the cloud, or be replicated in both locations. Figure 5 shows the decision tree.

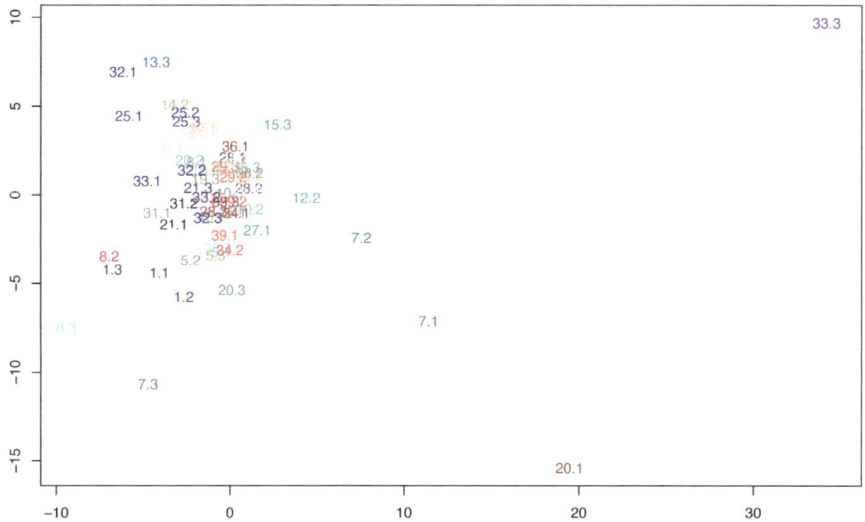

Fig. 6 A set of design alternatives: each individual generated three alternatives: for example, 7.1, 7.2, 7.3. The designs were compared based on their component connections, and the points mapped into the figure using multi-dimensional scaling. More detail on the method is discussed in [6]

The tree becomes a way of exploring the space and documenting decisions: each design is constituted by the nodes traversed from the root on the left to the leaves on the right. Thus, the entire design space, all possible designs, is a tree of trees. That is, there are a finite number of choices that can made in the above tree—there are four ways program can be deployed, and independently four ways data can be deployed, and thus there are 16 possible design alternatives. The tree might be a useful discovery aid in enumerating possible designs.

There are other ways of thinking about exploring the total design space For example, consider all the variations that have already been made. We might be able to explore the space by looking at the relative differences between the alternative designs of an individual designer, or better yet, the designs of a crowd of designers working independently. If a design can be expressed as a network, then the distance between two designs is just the graph edit distance [20]: the number of edges that would need to be either added or subtracted from the set [6]. Then, such a set of distances can be visualized using multidimensional scaling [21]. We show an example of this process in Fig. 6.

The above figure is derived from the designs of a crowd: that is, designers working independently. Each had created three variations, and the graph can be used to see how much individuals vary among themselves, and where the designs seem to converge.

Often design is done collaboratively, with developers modifying other's work. For example, the website Scratch [22, 23] provides a way for children to remix, that is modify, each other's programs, and thereby teach each other programming in

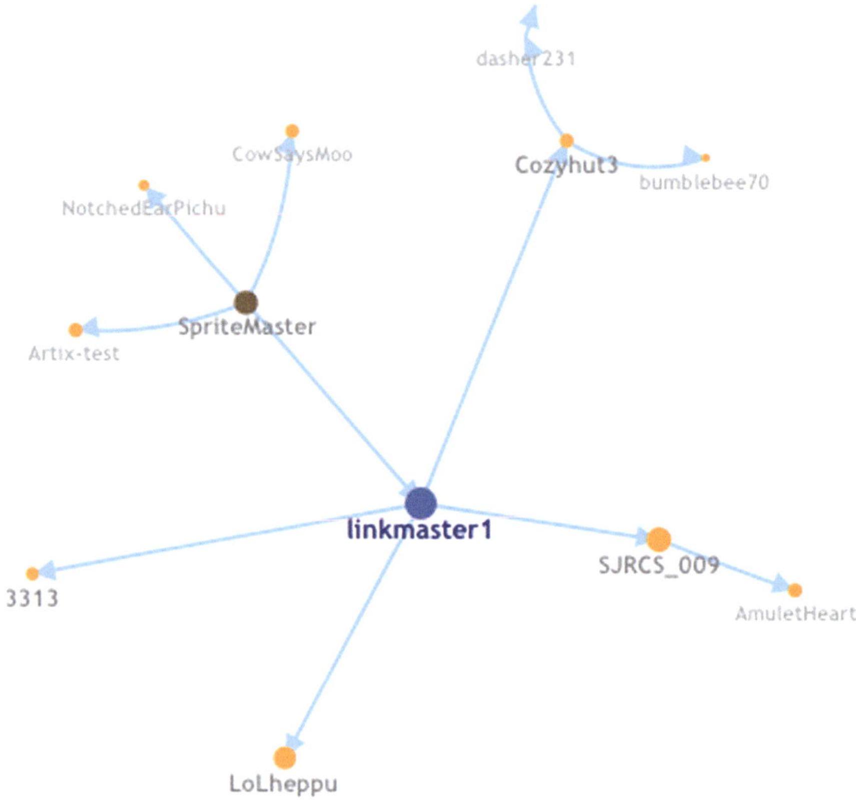

Fig. 7 A visualization showing the chain of programmers that have modified each other's code for a particular set of linked projects on the website Scratch: http://scratch.mit.edu/projects/SpriteMaster/1054710, as of 5/17/2010

the process of creating animations and games. In order to encourage remixing, the developers of Scratch have added a visualization, based on the history of remixes. Clicking on a node recenters the graph and emphasizes the local neighborhood of remixes [24]: the result is shown in Fig. 7. This diagram also gives a sense of both the design tree for a project, and the overall design space, by which variations have led to. One can imagine combining the methods shown in Figs. 6 and 7, by showing not only who modified a project, but how much the project was modified.

Evaluation Space

Designs not only need to be generated, they also need to be evaluated. There are often both a set of requirements with a design activity, and an overall set of criteria that create a space in which each design can be evaluated. For example, a word

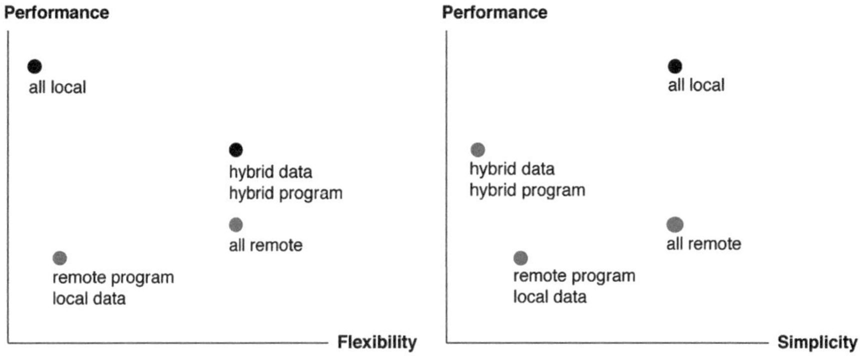

Fig. 8 On the left, **a** an evaluation of four designs from the design space of Fig. 5, based on the criteria of Performance and Flexibility. On the right, **b** these same designs evaluated on Performance and Simplicity

processor will be required to fulfill a long list of requirement relating to editing, file saving, and formatting. Assuming these can be fulfilled, there are set of general criteria that often determine the overall effectiveness of the system. For instance, the 16 designs that can be generated from Fig. 5 each can be evaluated in relation to performance and flexibility; Fig. 8a shows several designs mapped into this evaluation space.

The graph shows there is a tradeoff: local systems will perform better, but are less flexible. They are less flexible because users can't reach over the network and retrieve a file the same way they can with a cloud-based application. The designs shown as gray dots are relatively worse, because they perform no better along any dimension than the systems shown as black dots. In particular, a remote program that works off of a file on my local workstation is worse with respect to both performance and flexibility than an all local system or a hybrid system, in which both data and programming are distributed onto local and remote machines.

Designers fight over the criteria to be used in evaluation [25]. Someone who has learned from past experience that simplicity is an important systems virtue might substitute this criteria for flexibility, and would then choose a local solution, as shown in Fig. 8b. Furthermore, Brooks [3] points out that sometimes new criteria are discovered as part of the design process, and so there is often a shuttling back and forth between evaluation and generation, as ideas are generated, evaluated, and new solutions are sought that fill out parts of the design space. Criteria are sometimes added: for example, a team of programmers may decide that performance, flexibility and simplicity are all important. The designers then alternate between exploring the design space by making different choices in the tree shown in Fig. 5, and evaluating the solution in a three dimensional evaluation space, the two projections of which are shown in Fig. 8.

Not all designers work so rationally, and it is an open question how expert designers who claim they work intuitively successfully find designs that satisfy design criteria. There is, however, a class of designers that are by definition systematic:

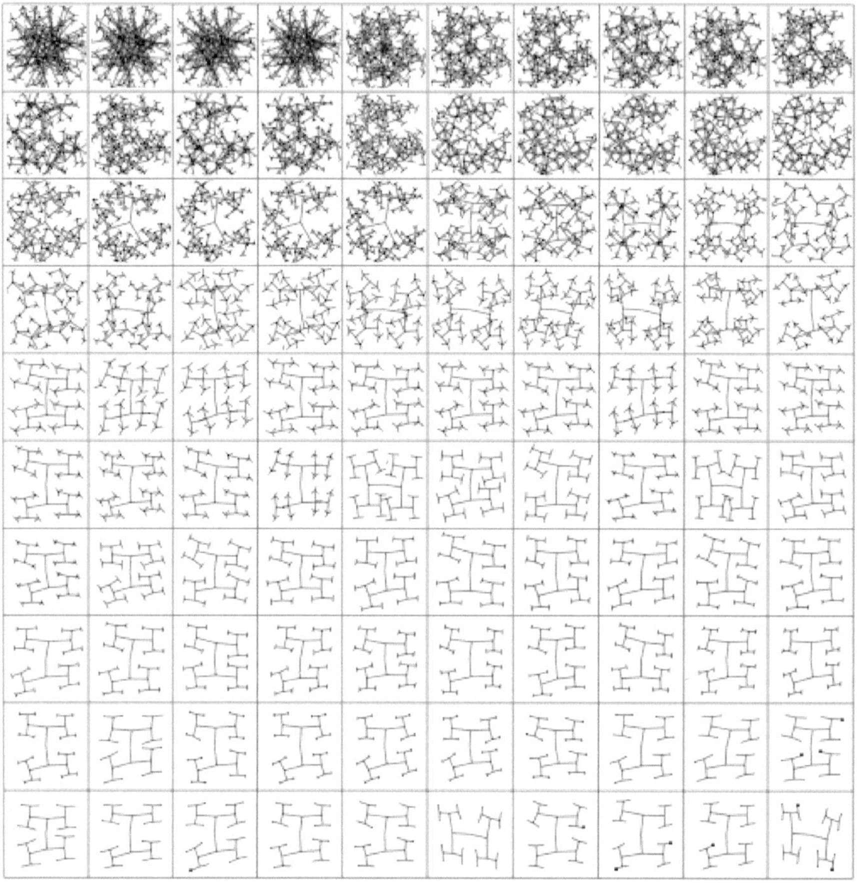

Fig. 9 A genetic algorithms exploration of a sensor network design space. Each cell is a point in the Pareto-Optimal set. The shape that optimizes the material used, in the bottom right corner, is an example of an H-Tree, re-discovered by the algorithm, but previously used in several areas, including VLSI design [31]

machines. In fields from architecture to engineering [26], automated systems explore design spaces, seeking optimal solutions: The output of such systems can look remarkably creative [27]. A range of techniques often used to explore design space are called meta-heuristic multi-objective optimization techniques [28]; for example, genetic algorithms combine solutions to create new ones, and these solutions are then considered in a multi-dimensional evaluation space so that a full range of alternatives are found. Such techniques are sometimes applied to computer programs themselves: that is, new software programs are generated automatically that fulfill specific design objectives [29]. Genetic algorithm techniques have also been applied to software architecture [30]. Figure 9 shows such a technique applied to a sensor network problem. The two criteria were: first, to minimize the cost of the network, as measured by the total amount of material used, and second, to maximize the amount of coverage. The 100 solutions shown in Fig. 9 are all optimal, meaning that there are no other solutions in the design space better than them, assuming all weights of the two criteria are equally valid.

When problems have a spatial aspect, then the chances of a correspondence between design space and evaluation space are increased. That is, a small change in the design space will generally result in a small change in evaluation space, and finding this correspondence can provide insights into the design process [32]. But in the case of an information systems design that involves a network, small changes in the network will often break the topology. Thus, a more indirect way of encoding the system is needed: for example, mapping each potential network onto a permutation [33]. Such indirect ways are useful in that they guarantee that any network considered will fulfill the systems requirements. However, our ability to reason spatially about the process becomes difficult or impossible, and the exploration of a permutation space may require large amounts of computational power. An open question is whether or not the intuitions of an expert design might suggest other ways of encoding and traversing the highly abstract topological spaces of information systems, perhaps through spatial reasoning. Such reasoning may exist even in the design of abstract systems, because simple spatial structures underlie many common cognitive tasks [34].

There is a non-automated approach that is showing promise. Crowdsourcing marketplaces (for example, [35]) make it possible to divide design tasks into small parts and allow thousands of human participants to engage in design activity. Can groups of independent designers tackle scale design problems? We saw before that humans can be used to generate individual alternatives that together may traverse a swath of the design space [6]. The crowd can also be used at a higher level to establish a correspondence between common situations and common technical mechanisms that are useful in such situations. Figure 10 shows the consensus of 30 designers about which technical mechanisms apply to a set of common situations [36].

When is the crowd better than an automated approach? When is close-knit team of designers better than a crowd? These are areas for exploration. It is possible that difficult problems may yield to a combination of traditional and new approaches to design; for example, close-knit team processes augmented by computational methods that perform evaluation, or crowd-based processes feeding unfinished ideas to expert designers for evaluation and refinement.

Concluding Thoughts

Design of information systems involves grappling with a set of abstract spaces. There is little visible in an information system, and much of the system is dynamic, transient. So the designer needs ways to get a handle on the system. Geography is important in few a situations, but in most situations connections are much more revealing. Therefore designers spend most of their time constructing networks that

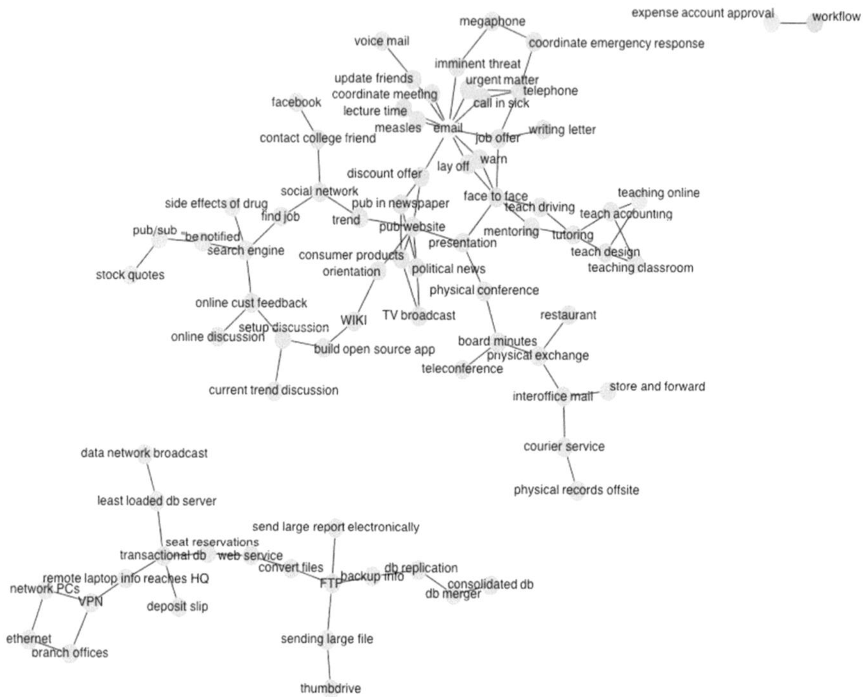

Fig. 10 The consensus evaluation of a set of designers asked to match technical mechanisms to common situations

together can describe the behavior of the system so that it can be constructed. Decisions are made, and the end result of the decisions are alternative designs. These many designs together form a design space, and variations in this space can be generated intuitively, or systematically by making different decisions in a design tree. But these choices need to be evaluated: once evaluated, the designer often moves back into generation mode, trying to find new solutions that explore a desirable part of the design space.

The design process is not always systematic, nor is it always conscious. Simon's rational decision approach [37] was critiqued by Winograd and Cross, among others [5, 38]. Still, much of what Simon said still underlies the current conversations about design science [39–41]. What has been tempered in current conversation is the belief in universal approaches and solutions. Domains are distinct, situations are different, and the design process itself is political [42]. We have dampened our enthusiasm for a proscriptive sequence of design activity, because we know that new requirements will be uncovered as the process proceeds [3, 43]. Yet Simon pointed out the importance of visualization, and we are even more convinced today of its importance in design [1, 2]. Fixed diagrammatic conventions, as in [7], are useful because they allow common communication, but we don't fully understanding how well these representations are understood by practitioners, whether alternative ones would be better, and how completely these representations cover the many abstract spaces that that need visual expression.

We know information systems design is both visual and verbal, and that abstraction is an important prerequisite to the production of novelty—generally, and in information systems design [44]. We know it is easy for novice designers to become confused by even simple abstract diagrams [11–12]. We also know that designers are inventive, creating hybrid representations to apply to particular situations [6, 8]. As a field, we are still in the process of learning how to guide the novice, and augment the expert, by providing appropriate tools and techniques. Looking to the future, new programming languages geared toward children are helping create communities of computationally fluent youngsters [22]. The children's collective community emulates the adult open source community, and both are examples of the growth of peer production in many facets of creative work [45]. Alongside this growth in human capacity is the growth of machine capacity, in clusters, desktops, laptops, tablets, and phones. Thus we anticipate fast increase in our collective cognitive and computational capacity to design information systems. Representations that integrate individual, team, crowd, and machine may be the levers of distributed cognition.

Acknowledgements This work was supported by grants from the National Science Foundation, IIS-0725223, IIS-0855995, and IIS-0968561.

References

1. Tversky B, Suwa M (2009) Thinking with sketches. In: Markman AB, Wood KL (eds) Tools for innovation. Oxford University, Oxford
2. Visser W (2006) The cognitive artifacts of designing. Erlbaum, Mahwah
3. Brooks FP (2010) The design of design: essays from a computer scientist. Addison-Wesley, Upper Saddle River
4. Shaw M, Garlan D (1996) Software architecture: perspectives on an emerging discipline. Prentice Hall, Upper Saddle River
5. Winograd T (1996) Bringing design to software. ACM Addison-Wesley, New York
6. Nickerson JV, Corter JE, Tversky B, Zahner D, Rho YJ (2008) The spatial nature of thought: understanding information systems design through diagrams. Proceedings of the International Conference on Information Systems, Paris, France
7. Booch G, Rumbaugh J, Jacobson I (2005) The unified modeling language user guide. Addison-Wesley, Upper Saddle River
8. Nickerson JV (2005) The meaning of arrows: diagrams and other facets in system sciences literature. 38th Hawaii International Conference on System Sciences
9. Hutchinson J (1989) Netscal: a network scaling algorithm for nonsymmetric proximity data. Psychometrika 54(1):25–51
10. Nickerson JV, Corter JE (2009) Clarity from confusion: using intended interactions to design information systems. Proceedings of the Fifteenth Americas Conference on Information Systems
11. Nickerson JV, Tversky B, Corter JE, Yu L, Mason D (2010) Thinking with networks. Proceedings of the 32nd Annual Conference of the Cognitive Science Society

12. Nickerson JV, Corter JE, Tversky B, Zahner D, Rho YJ (2008) Diagrams as tools in the design of information systems. In: Gero JS, Goel A (eds) Design computing and cognition '08. Springer, Dordrecht
13. Corter JE, Rho Y, Zahner D, Nickerson JV, Tversky B (2009) Bugs and biases: diagnosing misconceptions in the understanding of diagrams. Proceedings of the 31st Annual Conference of the Cognitive Science Society
14. Iivari J (1996) Why are CASE tools not used? Commun ACM 39(10):94–103
15. Schön DA (1983) The reflective practitioner: how professionals think in action. Basic Books, New York
16. Redmiles D, Nakakoji K (2004) Supporting reflective practitioners. Proceedings of the 26th International Conference on Software Engineering (ICSE'04) pp 688–690
17. Cottingham DN (2008) Vehicular wireless communication Ph.D. dissertation, University of Cambridge Computer Laboratory, September 2008
18. Nickerson JV (2005) A concept of communication distance and its application to six situations in mobile environments. IEEE Trans Mob Comput 4(5):409–419
19. Kunz W, Rittel H (1970) Issues as elements of information systems. Working paper No. 131, Studiengruppe für Systemforschung, Heidelberg, Germany, July, 20
20. Sanfeliu A, Fu KS (1983) A distance measure between attributed relational graphs for pattern recognition. IEEE Trans Syst Man Cybern 13:353–362
21. Shepard RN (1962) The analysis of proximities: multidimensional scaling with an unknown distance function. Psychometrika 27:125–140
22. Resnick R, Maloney J, Monroy-Hernández A, Rusk N, Eastmond E, Brennan K, Millner A, Rosenbaum E, Silver J, Silverman B, Kafai Y (2009) Scratch: programming for all. Commun ACM 52(11):60–67
23. Monroy-Hernández A, Resnick M (2008) Empowering kids to create and share programmable media. Interactions 15(2):50–53
24. Chen R, Monroy-Hernandez A (2010) Personal communication.
25. Nickerson JV, Yu L (2010) There's actually a car, perspective taking and evaluation in software-intensive systems design conversations, NSF Workshop on Studying Professional Software Design, Irvine.
26. Gero JS, Maher ML (eds) (1993) Modeling creativity and knowledge-based creative design. Erlbaum, Hillsdale
27. Bentley PJ, Corne DW (eds) (2002) Creative evolutionary systems. Morgan Kaufmann, San Francisco
28. Deb K (2001) Multi-objective optimization using evolutionary algorithms. Wiley, Chichester
29. Iba H, Paul TK, Hasegawa Y (2010) Applied genetic programming and machine learning. CRC, Boca Raton
30. Martens A, Koziolek H, Becker S, Reussner R (2010) Automatically improve software architecture models for performance, reliability, and cost using evolutionary algorithms. In Proceedings of the First Joint WOSP/SIPEW international Conference on Performance Engineering
31. Ullman JD (1984) Computational Aspects of VSLI. Computer Science Press, Rockville
32. Mackenzie CA, Gero JS (1987) Learning design rules from decisions and performances. Artif Intell Eng 2(1):2–10
33. Hodge LE, Whitaker RM, Hurley S (2006) Multiple objective optimization of bluetooth scatternets. J Comb Optim 11(1):43–57
34. Kemp C, Tenenbaum JB (2008) The discovery of structural form. Proc Natl Acad Sci 105(31):10687–10692
35. Amazon Mechanical Turk, Artificial Artificial Intelligence. https://www.mturk.com/mturk/welcome.
36. Nickerson JV, Zahner D, Corter JE, Tversky B, Yu L, Rho YJ (2009) Matching mechanisms to situations through the wisdom of the crowd. Proceedings of the International Conference on information systems
37. Simon HA (1969) The sciences of the artificial. MIT, Cambridge

38. Cross N (2006) Designerly ways of knowing. Springer, London
39. Gross MD (2003) How is a piece of software like a building? Toward general design theory and methods, in the National Science Foundation (NSF) Invitational Workshop on Science of Design, Software Intensive Systems, Airlie Center, Virginia
40. Hevner AR, March ST, Park J, Ram S (2004) Design science in information research. MIS Quarterly 28:75–105
41. Lee AS, Nickerson JV (2010) Theory as a case of design: lessons for design from the philosophy of science. Proceedings of the 43rd Annual Hawaii International Conference on systems science
42. Bergman M, King J, Lyytinen K (2002) Large scale requirements analysis revisited: the need for understanding the political ecology of requirements engineering. Requir Eng J 7(3):152–171
43. Gero JS (2003) Situated computing: a new paradigm for design computing. In Choutgrajank A, Charoenslip E, Keatruangkamala K, Nakapan W (eds), CAADRIA03, Rangsit University, Bangkok, pp 579–587
44. Zahner D, Nickerson JV, Tversky B, Corter JE, Ma J (2010) A fix for fixation? Re-representing and abstracting as creative processes in the design of information systems, Artificial Intelligence in Engineering Design, Analysis and Manufacturing, Maher M, Kim YS, Bonnardel N (eds) 24, 2.
45. Benkler Y (2006) Wealth of networks: how social production transforms markets and freedom. Yale University, New Haven

The Theoretical Framework for Creative Visual Thinking

Ewa Grabska

Introduction

Although designers during design process can form mental images in their heads, the majority of them do much better when those images are out in their internal worlds, in the form of drawings on paper or monitor screen. Therefore the main work of creative visual design is done through the dialogue with graphical outputs, which is very difficult to handle in a formal way. It is often simplified and perceived as *seeing* and *doing*. Understanding of this dialogue helps us to learn about constructive power of perception that has profound implications for creativity design [1]. This paper is an attempt to present a formal coherent framework for creative thinking which includes this dialogue, where human visual perception is treated as a dynamic process. The framework will be called a *creative visualization system*.

The paper deals also with other important aspects of creative design: *constructive perception* and *semantic convention*. The former is based both on external graphical information and on the constructive mental process. The latter governs the process of designing in which the designer creates drawings. In this paper the constructive perception is treated as a composition of a perceptual action (normal seeing) and a functional one (visual intelligence and imagination). Whereas the semantic convention is defined as a binary relation between constraints on graphical representations and constraints determined by designer's requirements.

A manner in which the designer thinks about design problems with graphical outputs is the next essential aspect for creative design. There are two major categories of thinking: *divergent* and *convergent* [2]. Divergent thinking is imaginative and intuitive. This ability has been associated with skill in the arts and it has been interpreted as an open-ended approach seeking alternative. This types of thinking is based on abduction and it is typical in an inventive design when a number of unknown design concepts is sought. Whereas convergent thinking is associated with science. It requires deductive and interpolative skills. The convergent thinking is

E. Grabska (✉)
Jagiellonian University, Kraków, Poland
e-mail: uigrabsk@cyf-kr.edu.pl

J. S. Gero (ed.), *Studying Visual and Spatial Reasoning for Design Creativity*,
DOI 10.1007/978-94-017-9297-4_3

logical and rational. Taken as a whole, design is a divergent task. However, during the process of creative design good designers are able to develop and maintain several lines of thought, both convergent and divergent. The proposed framework, in the form of the creative visualization system, enables one to handle both types of thinking in a formal way [3, 4].

Design Actions and Domains

Recent frameworks for creative design focus on dynamic character of the context in which designing takes place. The process of design is described by means of actions defined with the help of both an *internal world* and an *external world*. The internal world is built up inside the designer in terms of concepts and perceptions, while the external world is composed of graphical representations outside the designer [5].

Actions

During the design process designer's actions connect his/her internal and external worlds. The following four types of actions are distinguished [6]:

- *physical* actions—drawing, copying and erasing elements of graphical outputs,
- *conceptual* actions—finding requirements and setting design goals,
- *perceptual* actions—discovering visual features of drawings, such as for instance spatial relations between visual elements, and
- *functional* actions—associating meaning with features discovered in the perceptual actions and valuation of drawings.

This classification of actions makes designer's dialogue with drawings easier to characterize.

Domains of Creative Visualization

To unify the description of the considered designer's dialogue three components are distinguished (see: Fig. 1):

- a domain $\boldsymbol{D}_T$ of *design tasks* related to formulation of design problems in terms of requirements,
- a domain $\boldsymbol{D}_A$ of *physical design actions*, and
- a domain $\boldsymbol{D}_V$ of *visualization sites,* which consists of drawings along with a surface on which they are drawn.

Design process presented in the context of these three domains can be described with the use of two environments: an internal and external worlds.

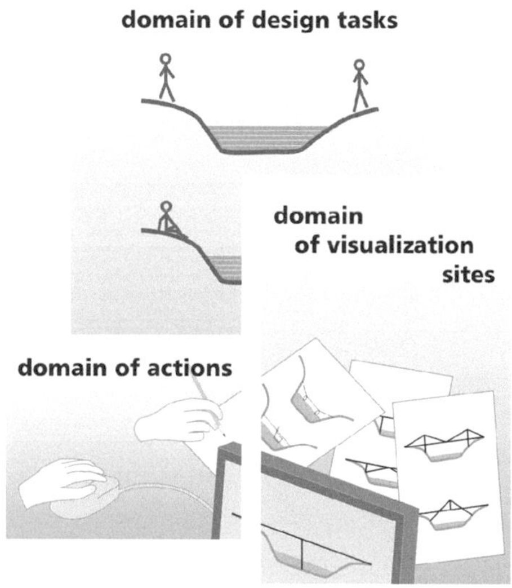

Fig. 1 The do-mains of creative visualization

The domain of design tasks is related to the internal world and is modified during the design process. At the beginning it contains only initial requirements, while later the devised requirements are added. When the designer determines design goals and requirements then conceptual actions are undertaken. Physical design actions of the domain of actions are related to the external world. The domain of visualization sites in the external world is associated with undertaking perceptional actions in the internal world. The remaining actions constructed in designer's brain are induced by perception which strongly determines the course of designing.

Types of Designer's Dialogue

Our study on creative design starts with the presentation of two types of designer's dialogue described in the context of the domain of tasks, the domain of visualization sites and the domain of actions. First, sketching, which commences design process is discussed and then the process of computer aided visual designing.

Sketching

Sketching is one of the best ways to absorb design ideas (Fig. 2). During drawing pictures there exists the need to pass an idea from mind to hand and to eye [2]. The mind in which the designer tries to formulate of design problems is related to the domain of design tasks. The hand symbolizes the domain of actions consisting in

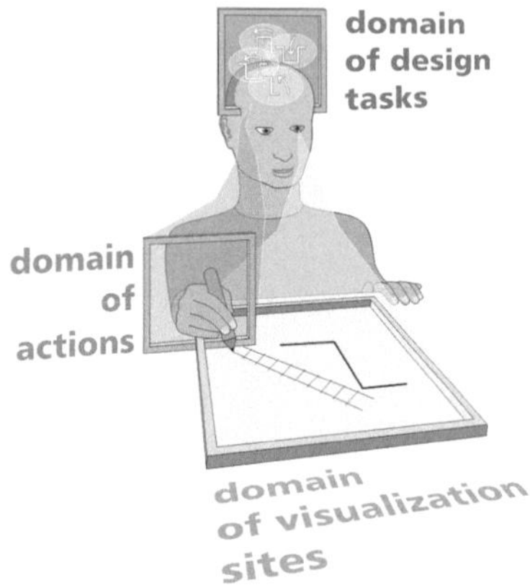

Fig. 2 The process of sketching

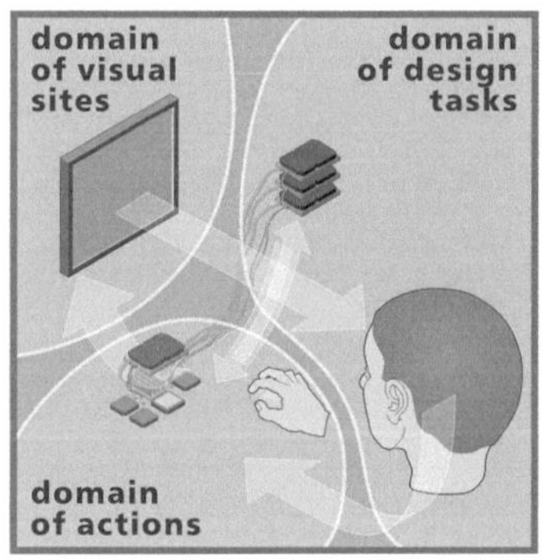

Fig. 3 Computer aided visual design

making drawings with a pencil or a pen. The eye is associated with designer's visual perception of the drawings on sheets of papers belonging to the domain of visual sites. In actions of drawing stimulation and perception are tightly intertwined [7].

Computer Aided Visual Design

The second example presents computer aided design process with the use of visualization. The domain of design tasks is related to formulation of design problems by the designer with the aid of computer system (see: Fig. 3).

Actions are aided by computer tools, for instance graphical editors or functional-structure editors. The domain of actions is about drawings with the use of a mouse, a tablet or a programming language. The results of performing physical actions are displayed on the monitor screen which belongs to the domain of visualization sites. Expression drawings on the monitor screen with the use CAD—tools has to respect the semantic convention between designer's requirements and requirements on graphical outputs.

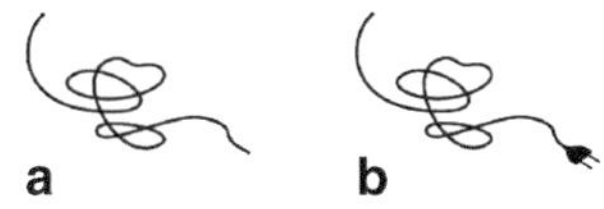

Fig. 4. **a** A random irregular scribble, **b** The same scribble with addition of a plug shape becomes a flex

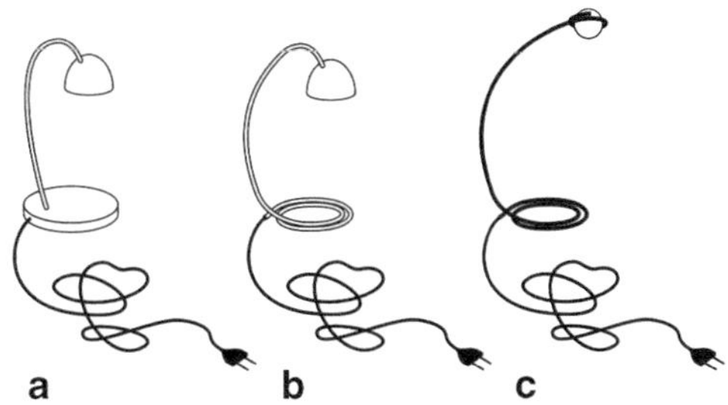

Fig. 5 The phases of designing lamp

The Fundamentals of Creative Design

Constructive perception and semantic convention are fundamental aspects of the creative design which can be described in the context of the considered three domains.

Constructive Perception

A composition of a perceptual action and a functional one can be seen as *constructive perception* which combines normal seeing with visual imagination [5]. The constructive perception connects the domain of visualization sites with the domain of design tasks, i.e., the external world with the internal one. Moreover it can stimulate a conceptual action in the form of a devised requirement added to the domain of design tasks.

Example 1 The following exercise allows one to understand how the constructive perception of lines with one simple shape visual added can be used as part of a creative design process (compare: [7], pp. 216–217). Let us make a simple scribble without thinking about representing anything (Fig. 4a). Let us add a shape of a plug to transform this meaningless scribble into an electric wire (a flex) shown in Fig. 4b. An irregular looping line can be seen as a looping flex.

This exercise can be a source of inspiration for drawing a piece of equipment, for instance a lamp connected through the flex to an electricity supply. The first attempt of adding a classical lamp is shown in Fig. 5a. After the perceptual action searching for similarities between design elements the lamp shape is modified (see: Fig. 5b). The final solution is presented in Fig. 5c.

Semantic Convention

Designer's requirements can be treated as constraints on expected design solutions. Forms of visual constraints on the drawings in the domain of visualization sites are different from forms in which the designer expresses requirements related to the design tasks. When taking physical actions the designer encode information about the object being designed in the fictional depicted world. He/she also deals with visual organization of the drawing, which includes form, proportion, line, shape and so on. The correspondence between constraints on drawings and constraints on designer's requirements determines a *semantic convention* which relates constraints on graphical representations to designer requirements.

A System of Creative Visualization

In this section creative design is described in the framework of the system of visual design. The system has three major components, including a domain of design tasks related to the imposed design problem and a domain of actions connected with the process of drawing in terms of physical actions, and a domain of visualization sites being about perception. Each of the three domains is defined with the use of the notation of classification, i.e., is described by a pair containing a set of objects to be classified and a set of types used to classify the objects [8].

The Domain of Design Tasks

Design task can be posed in terms of the design solution expected. A design solution describes a certain situation which is classified by types.

A domain of **design tasks** *is* a triple $D_T = (T, \Sigma_T, |\!-_T)$ consisting of a set T of objects to be classified, called *design situations* of D_T, a set Σ_T of objects used to classify the situations, called *types* of D_T, and the relation $|\!-_T$ contained in $T \, x \, \Sigma_T$, called a *belonging* relation.

If a design situation $t \in T$ is classified as being of type $\sigma \in \Sigma_T$, we write $t \, |\!-_T \sigma$ and say *t belongs to* σ Design situations are classified by design requirements in the form of expressions or sentences of the *propositional logic.*

Example 2 Let us consider the very simple example of the domain of design tasks. Its role is only to provide insight into the nature of this domain.

Let T be a set of decorative elements with rotational symmetry.

$\Sigma_T = \{\sigma_1, \sigma_2\}$ contains two types in the form of the following sentences:

σ1: design has four-fold rotational symmetry,

σ2: design is in the shape of a circle.

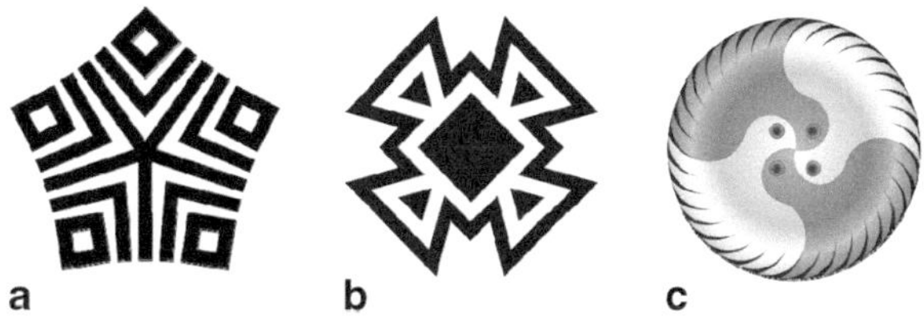

Fig. 6 Three de-sign situations

Three design situations of T shown in Fig. 6 are classified by two requirements expressed by sentences σ_1 and σ_2. Element t_1 in Fig. 6a is not either of type σ_1 or of type σ_2, whereas element t_2 in Fig. 6b is of type σ_1. Fig. 6c presents element t_3 which is both of type σ_1 and of type σ_2.

The Domain of Visualization Sites

Drawings are graphical representations which designers use in their external world. An arbitrary surface on which a drawing is made along with this drawing is called a *visualization site*. Two different drawings on the same surface, e.g., on the sheet of paper or on the monitor screen determine two different visualization sites. A visualization site is itself a situation in the external world, and as such it belongs to its own domain of classification, just as a design task situation does.

A domain of **visualization sites** is a triple $D_V=(V, \Omega_V, |-_V)$ consisting of a set V of objects to be classified, called *visualization sites* of D_V, a set Ω_V of *types* used to classify the situations, and the *belonging relation* $|-_V$ contained in $V x \Omega_V$.

Visual perception plays an essential role for the domain of visualization sites. If a visualization site v is used to find a design solution t then *v signals t* and we write $v \rightarrow t$. *Signaling,* denoted by $\rightarrow$ is a binary relation from V to T.

As it has been considered, the semantic convention relates constraints on graphical representations to designer requirements. Therefore, types of Ω_V that classify visual sites must be related to types Σ_T that classify design situations. The semantic convention is expressed as a binary relation => from Ω_V to Σ_T. The constructive perception is described by two relations, signaling and semantic convention, which together form a mapping from the domain of visualization sites to the domain of design tasks. The designer discovers information σ related to the design situation t from a visual site v only if $v \rightarrow t$ and there exists ω such that $v |-_V \omega$ and $\omega => \sigma$.

Example 3 Let us come back to the Example 2 and assume that after drawing the flex shown in Fig. 4b the process of designing a lamp has been continued with the use of the computer with the visualization site being a monitor screen. Usually the design process is started with an empty monitor screen (initial visualization site). Each drawing (Fig. 7) being a step of a design solution leading to a final drawing is treated as a different visualization site and constitutes a different design situation.

Fig. 7 The visual sites signaling the task of designing a lamp

The Physical Design Actions Domain

The last of the three domains describes. physical actions treated as a certain kind of events in the external world that start with an initial situation and result in another situation.

A domain of **physical design actions** $D_A = (A, \Delta_A, |\!-A)$ consists of a set A of physical actions to be classified, a set of types used to classify the situations, and the *belonging relation* $|\!-A$ contained in $A \times \Delta_A$.

If situation $a \in A$ is classified as being of type $\delta \in \Delta_A$ we write $a \; |\!-A \; \delta$ and say *a belongs to* δ.

The System

Summing up the discussion on the three design domains and relations between them, a visual design system is defined.

A **system of creative visualization** is a 5-tuple

$$\boldsymbol{CV} = (D_T, D_V, D_A, \Rightarrow, \rightarrow), \text{where:}$$

- D_T is a domain of design tasks,
- D_V is a domain of visualization sites,
- D_A is a domain of physical design actions,
- $\Rightarrow$ is a semantic convention, and
- $\rightarrow$ is a signaling relation.

The system $\boldsymbol{CV}$ allows one to define essential elements of creative visual designing.

Creative Visual Designing

Studies of designers, artists, and scientists have identified some common elements of creative visual thinking. Looking at the stages of the creative process from a generalized point of view, four steps are distinguished [1]:

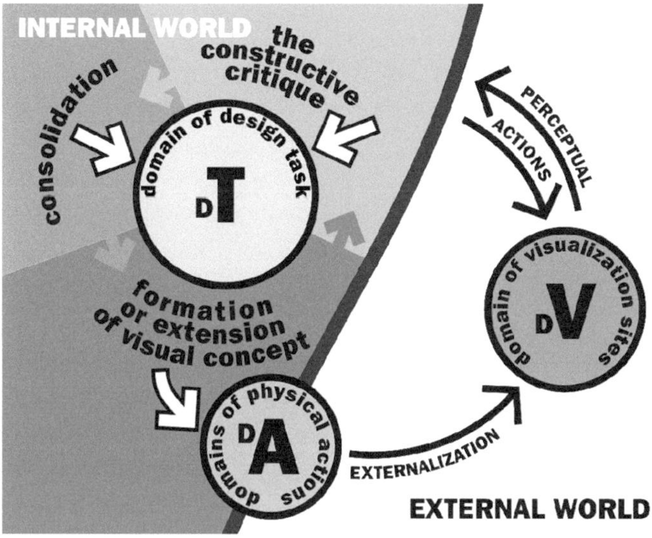

Fig. 8 Creative visual thinking

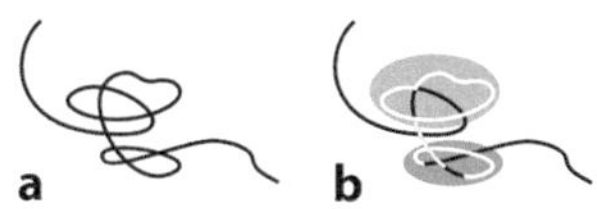

Fig. 9 Emergent shapes

1. formation of the visual concept—depending on the application, a free or stereotyped graphic idea is proposed,
2. externalization—a visualization site is created,
3. the constructive critique—constructive perception is used,
4. consolidation and extension—new requirements are formulated on the base of conceptual actions.

Figure 8 presents creative visual thinking in the framework of the system of creative visualization. The visual concept is forming first by means of types $\sigma_1, \ldots, \sigma_n$ expressing design requirements of the domain D_T. Then these types are transformed into types of actions $\delta_1, \ldots, \delta_k$ allowing one to determine a sequence of physical actions $a_1, \ldots, a_m$ of the domain D_A for drawing an appropriate graphical representation on the visualization site.

One of the element of the process of the constructive critique may be the *emergence* of new shapes. The designer discovers a new shape (which had not been consciously constructed) related to the design situation from a visual site.

Example 4 Let us consider the scribble shown in Fig. 9a. An example of emergent shape drawn white line is presented in Fig. 9b. The perceptual action allows the designer to notice this shape, while the functional action associates it with shapes

Fig. 10 Designing a lamp

of a lamp. This association becomes a new inspiration in creating a form of a lamp (Fig. 10) and enables the designer to formulate a devised requirement σ^* (a new type of Σ_T).

Let ***CV*** *be* a system of creative visualization. We say that **emergence** occurs in ***CV*** if:

- The sequence of physical actions realizes a fact $\omega^* \in \Omega_V$ on the visualization site.
- According to the semantic convention ($\Rightarrow$) an element ω^* of Ω_V can be transformed into a new type σ^* of Σ_T

Occurring emergence is an example of the convergent thought. The process of designing shown in Fig. 5 is an example of the divergent thought. Both convergent thought and divergent thought stimulate consolidation and extension in the process of creative visual thinking.

Conclusions

Nowadays, visual designer environment plays an essential role in creative visual designing. The proposed system of creative visualization shows different aspect of visual design process. To develop this system, which is necessary, in the one hand for deeply understanding the fundamentals of creativity, and in the second hand to devise new visual tools, a higher level of abstraction had to be used. The new framework for creative design allows one to hold concepts from different disciplines (engineering and psychology) in a formal way and shows influence of different perspective on the design theory.

References

1. Ware C (2008) Visual thinking for design. Elsevier
2. Lawson B (2001) How designers think: the design process demystified. Butterworth Architecture. Oxford
3. Grabska E (2007) Computer-aided visual design (in polish). EXIT, Warszawa
4. Arciszewski T, Grabska E, Harrison C (2009) Visual thinking in inventive design: three perspective, (invited). Proceedings of the first international conference on soft computing in civil, structural, and environmental engineering. Madeira, Portugal
5. Gero JS, Kannengiesser U (2002) The situated function-behaviour—struc-ture framework. In: Gero JS (ed) Artificial intelligence in design'02. Kluwer, Dordrecht, pp 89–104

6. Suwa M, Gero JS, Purcell T (2000) Unexpected discoveries and S-invention of design requirements: important vehicles for a design process. Des Stud 21(6):539–567
7. Massironi M (2002) The psychology of graphic images: seeing, drawing, communicating. Lawrence Erlbaum
8. Barwise J, Seligman J (1997) The logic of distributed systems. Cambridge University Press

Sortal Grammars for Urban Design

Rudi Stouffs

Introduction

Grammar formalisms have been around for over 50 years and have found application in a wide variety of disciplines and domains, to name a few, natural language, architectural design, mechanical design, and syntactic pattern recognition. Grammar formalisms come in a large variety (e.g., [1–5]), requiring different representations of the objects being generated, and different interpretative mechanisms for this generation. Altering the representation may necessitate a rewrite of the interpretative mechanism, resulting in a redevelopment of the entire system. At the same time, all grammars share certain definitions and characteristics. Grammars are defined over an algebra of objects, U, that is closed under the operations of addition, +, and subtraction, −, and a set of transformations, F. In other words, if u and v are members of U, so too are $u + f(v)$ and $u - f(v)$ where f is a member of F. In addition, a match relation, ≤, on the algebra governs when an object occurs in another object under some transformation, that is, $f(u) \leq v$ whenever u occurs in v for some member f of F, if u and v are members of U.

Building on these commonalities, we consider a component-based approach for building grammar systems, utilizing a uniform characterization of grammars, but allowing for a variety of algebras, and match relations (or interpretative mechanisms) [6]. *Sortal* representations constitute the components for this approach. They implement a model for representations, termed *sorts*, that defines formal operations on *sorts* and recognizes formal relationships between *sorts* [7]. Each *sort* defines an algebra over its elements; formal compositions of *sorts* derive their algebraic properties from their component *sorts*. This algebraic framework makes *sortal* representations particularly suited for defining grammar formalisms. Provided a large variety of primitive *sorts* are defined, *sortal* representations can be conceived and built corresponding to almost any grammar formalisms.

R. Stouffs (✉)
Delft University of Technology, Delft, The Netherlands
e-mail: r.m.f.stouffs@tudelft.nl

J. S. Gero (ed.), *Studying Visual and Spatial Reasoning for Design Creativity*,
DOI 10.1007/978-94-017-9297-4_4

The need for varying grammar formalisms using varying representations is quite apparent in urban design. CAD systems are very powerful drawing tools and fit for design practice, also in urban design. On the other hand, GIS systems are very powerful systems for accessing large-scale urban data; hence they play an important role in urban planning as analytical tools. However, these tools were conceived as interactive maps and so they lack capacities for designing. Therefore, in urban design, the linking of GIS to CAD tools and representations becomes an important goal to allow designing directly on the GIS data.

For urban design and simulation, *sortal* grammars may include, among others, descriptive grammars, GIS-based set grammars, shape grammars and any combination thereof.

Sortal Representations

Stouffs [7] describes a semi-constructive algebraic formalism for design representations, termed *sorts*, that provides support for varying grammar formalisms. It presents a uniform approach for dealing with and manipulating data constructs and enables representations to be compared with respect to scope and coverage, and data to be converted automatically, accordingly. *Sorts* can be considered as hierarchical structures of properties, where each property specifies a data type; properties can be collected and a collection of one or more properties can be assigned as an attribute to another property. *Sorts* can also be considered as class structures, specifying either a single data type or a composition of other class structures.

Each *sort* has a behavioral specification assigned, governing how data entities combine and intersect, and what the result is of subtracting one data entity from another or from a collection of entities from the same sort. This behavioral specification is a prerequisite for the uniform handling of different and a priori unknown data structures and the effective exchange of data between various representations. The behavioral specification of a *sort* is based on a part relationship on the entities of this *sort*, with the *sortal* operations of addition, subtraction, and product defined in accordance to this part relationship. As such, a behavioral specification explicates the match relation (or interpretative mechanism) underlying a *sortal* algebra and grammar. The behavioral specification of a primitive *sort* forms part of the predefined template of this *sort*; composite *sorts* derive their behavioral specification from the component *sorts* in conformity with the compositional operation. In addition, a functional *sort* allows the specification of data (analysis) functions that automatically apply to *sortal* structures through tree traversal.

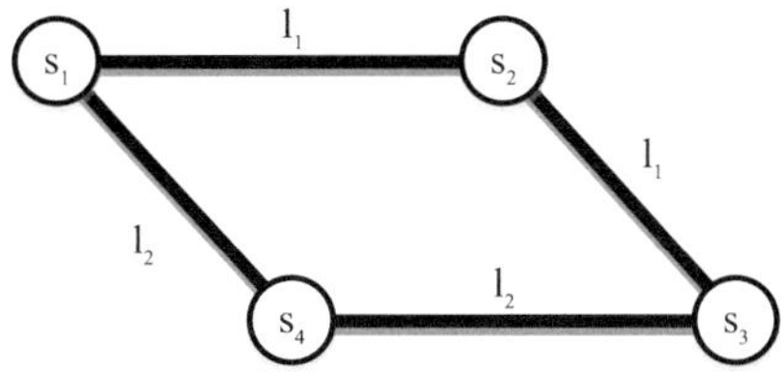

Fig. 1 A simple transportation network consisting of two lines and four stops

A Simple Example

Consider the following example: given a public transportation network, where the transportation nodes represent stations or stops and the edges represent transportation lines, how can we derive a transportation lines connectivity graph, where the nodes represent transportation lines and the edges exchanges between these lines? Basically, we are interested in knowing how many lines there are, which stations or stops are on which line, which lines connect to one another, etc., such that we can take into account the number of exchanges that might be necessary to get from one point to another. We assume that stations and stops have attribute information specifying the lines that stop here.

From a programming point of view, the derivation of a line connectivity graph from a transportation network or stop connectivity graph is not all that complex, but without proper programming knowledge, the task can still be very challenging. We show how one might approach this problem using *sortal* structures. First, we need to define the representational structure we will use as a starting point.

Figure 1 illustrates the data that may be present in the stop connectivity graph. We ignore the format in which the data may be provided, and instead consider the basic data entities that are required. Firstly, we need to represent the stops themselves, e.g., “s1”, “s2”, “s3” and “s4”. We can do so by their name, a string. We define a primitive *sort* with Label as *sortal* template:

```
sort stops : [Label];
form $stops = stops: { "s1", "s2", "s3", "s4"};
```

The first line defines the *sort* stops, with template Label. The second line defines an exemplary data form of *sort* stops and referenced by the *sortal* variable $stops. It defines a collection of stops or, more specifically, stop labels “s1” through “s4”. Similarly, we can represent the transportation lines that use a stop also as strings with *sortal* template Label:

```
sort lines : [Label];
form $lines = lines: { "l1", "l2" };
```

Finally, we need to represent the connectivity relations. For this, we use the Property template. Unlike other templates, the Property template requires two primitive *sorts* as arguments, and defines not one but two new primitive *sorts*:

```
  sort   (connections,   rev_connections)   :   [Property]
(stops, stops);
```

The two arguments define the representational structure for the tails and heads of the connectivity relationship. Since the Property template applies to directional relationships, we consider two resulting sorts: connections and rev_connections (reverse connections). An example is given below.

A complex representational structure is defined as a composition of primitive representational structures. *Sortal* structures offer us two compositional operators: an attribute operator, ^, specifying a subordinate, conjunctive relationship between *sortal* data, and a sum operator, +, specifying a co-ordinate, disjunctive relationship. Considering the room adjacency graph, we can define a corresponding *sortal* structure as follows:

```
  sort  input  :  stops  ^  (lines  +  connections  +
rev_connections);
```

Stops have lines, connectivity relationships and reverse connectivity relationships as attributes. A corresponding data form would be defined as:

```
form $input = input:
{ #me-stops-1 "s1"
  { (lines): { "l1" },
    (connections): { me-stops-2, me-stops-4 } },
  #me-rooms-2 "r2"
  { (lines): { "l1" },
    (connections): { me-stops-3 } },
  #me-rooms-3 "r3"
  { (lines): { "l1", "l2" },
    (connections): { me-stops-4 } },
  #me-rooms-4 "r4"
  { (lines): { "l2" } } };
```

Of course, this data form may be generated from the original data format, rather then specified in textual form. Especially, the connectivity relationships may be generated automatically from a sequentially-ordered list of stops on a transportation line. #me-stops-1 is a reference ID for "s1" that can be used later, in the form me-stops-1, to reference "s1" in an connectivity relationship from a different stop. The specification of reverse connectivity relationships is optional; the *sortal* interpreter will automatically generate these.

Similarly, we can define a representational structure for the output we need to produce. Consider the goal to group stops on the same line. For this, we can consider lines with stops as attributes; the stops themselves may still have (reverse) connectivity relationships as attributes:

```
  sort   output   :   lines   ^   stops   ^   (connections   +
rev_connections);
  form $output = output: $input;
```

The second line defines a variable of *sort* output with $input as data. Since $input is defined of sort input, the data must be converted to the new *sort*. This conversion is done automatically based on rules of semantic identity and syntactic similarity. The result is:

```
  form $output = output:
  { "l1"
    { #me-stops-1 "s1"
      { (connections): { me-stops-2, me-stops-4 } },
      #me-stops-2 "s2"
      { (connections): { me-stops-3 },
        (rev_connections): { me-stops-1 } } },
      #me-stops-3 "s3"
      { (connections): { me-stops-4 },
        (rev_connections): { me-stops-2 } } },
    "l2"
    { #me-stops-1 "s1"
      { (connections): { me-stops-2, me-stops-4 } },
      #me-stops-3 "s3"
      { (connections): { me-stops-4 },
        (rev_connections): { me-stops-2 } },
      #me-stops-4 "s4"
      { (rev_connections): { me-stops-1, me-stops-3 } }
} };
```

This is a collection of transportation lines, with for each line a list of stops (ordered alphabetically, rather than sequentially), with stop connectivity relationships (and reverse relationships). It does not yet constitute a transportation lines connectivity graph. For this, we need to define relationships (and reverse relationships) between transportation lines:

```
  sort (exchanges, rev_exchanges) : [Property] (lines,
lines);
```

We can now consider lines with exchange relationships (and reverse relationships); the lines may still have stops as attributes but for the automatic conversion of the relationships to take place, we must omit the stop connectivity relationships.

```
  sort graph : lines ^ (stops + exchanges +
rev_exchanges);
  form $graph = graph: $ouput;
```

The result will be:

```
  form $graph = graph:
  { #me-lines-1 "l1"
    { (stops): { "s1", "s2", "s3" },
      (exchanges): { me-lines-1, me-lines-2 },
      (rev_exchanges): { me-lines-1 } },
    #me-lines-2 "l2"
    { (stops): { "s1", "s3", "s4" },
      (exchanges): { me-lines-2 },
      (rev_exchanges): { me-lines-1, me-lines-2 } } };
```

Using functional entities integrated in the representational structures, we can also calculate the number of lines, the number of stops per line, etc. For this, we define a new primitive *sort* with Function as *sortal* template, and define a representational structure of counting functions with lines as attribute, where the lines themselves have stops as attributes, though we ignore any relationships:

```
  sort counts : [Function];
  sort number_of_lines: counts ^ lines ^ stops;
  // func count(x) = c : {c(0) = 0, c(+1) = c + 1};
  ind $count = number_of_lines: count(lines.length)
$output;
```

The last line defines a data form as an individual (a single data entity, not a collection of data entities or individuals) contained in the variable $count of *sort* number_of_lines. This individual consists of the count function applied to the length property of the *sort* lines. The function count is pre-defined in the *sortal* interpreter but, otherwise, could be specified as shown in the comment (preceded by '//'). A function always applies to the property of a *sort*. In this case, the exact property doesn't matter as its value is not actually used in the calculation of the result of the count function. The length property of a *sort* with Label as *sortal* template specifies the length—the number of characters—of the corresponding label. The result is:

```
  ind $count = number_of_lines: count(lines.length) =
2.0
  { "l1"
    { #me-stops-1 "s1",
      #me-stops-2 "s2",
      #me-stops-3 "s3" },
    "l2"
    { #me-stops-1 "s1",
      #me-stops-3 "s3",
      #me-stops-4 "s4" } };
```

Similarly, in order to calculate the number of stops per line, we can apply the function count to the length property of the *sort* stops. We reuse the *sort* number_of_lines for now.

```
  ind $count = number_of_lines: count(stops.length)
$output;
```

However, the result will be incorrect as stops belonging to multiple lines will be counted as many times. In order to correct the result, we need to alter the location of the count function in the representational structure to be an attribute of the *sort* lines. We can achieve this simply by creating a new *sort* and relying on the automatic conversion of one data form (or individual) into another.

```
    sort number_of_stops: lines ^ counts ^ stops;
    form $stops_per_line = number_of_stops: $count;
```

The result is:

```
form $stops_per_line = number_of_stops:
{ "l1"
  { 3.0
    { #me-stops-1 "s1",
      #me-stops-2 "s2",
      #me-stops-3 "s3" } },
  "l2"
  { 3.0
    { #me-stops-1 "s1",
      #me-stops-3 "s3",
      #me-stops-4 "s4" } } };
```

Sortal Grammars

Grammars are formal devices for specifying languages. A grammar defines a language as the set of all objects generated by the grammar, where each generation starts with an initial object and uses rules to achieve an object that contains only elements from a terminal vocabulary. A rewriting rule has the form *lhs* → *rhs*; *lhs* specifies the similar object to be recognized, *rhs* specifies the manipulation leading to the resulting object. A rule applies to a particular object if the *lhs* of the rule 'matches' a part of the object under some allowable transformation. Rule application consists of replacing the matching part by the *rhs* of the rule under the same transformation. In other words, when applying a rule $a \rightarrow b$ to an object s under a transformation f such that $f(a) \leq s$, rule application replaces $f(a)$ in s by $f(b)$ and produces the shape $s—f(a)+f(b)$. The set F of valid transformations is dependent on the object type. In the case of geometric entities, the set of valid transformations, commonly, is the set of all Euclidean transformations, which comprise translations, rotations and reflections, augmented with uniform scaling. In the case of textual entities, or labels, case transformations of the constituent letters may constitute valid transformations.

The central problem in implementing grammars is the matching problem, that of determining the transformation under which the match relation holds for the *lhs*. Clearly, this problem depends on the representation of the elements of the algebra. *Sorts* offer a representational flexibility where each *sort* additionally specifies its own match relation as a part of its behavior. For a given *sort*, a rule can be specified as a composition of two data forms, a *lhs* and a *rhs*. This rule applies to any particular data form if the *lhs* of a rule is a part of the data form under any applicable transformation f, corresponding to the behavioral specification of the data form's *sort*. Rule application results in the subtraction of $f(lhs)$ from the data form, followed by the addition of $f(rhs)$ to the result. Both operations are defined as part of the behavioral specification of a *sort*.

As composite *sorts* derive their behavior from their component *sorts*, the technical difficulties of implementing the matching problem only apply once for each primitive *sort*. As the part relationship can be applied to all kinds of data types, recognition algorithms can easily be extended to deal with arbitrary data representations, considering a proper definition of what constitutes a transformation. Correspondingly, primitive *sorts* can be developed, distributed, and adopted by users without any need for reconfiguring the system. At the same time, the appropriateness of a given grammar formalism for a given problem can easily be tested, the formalism correspondingly adapted, and existing grammar formalisms can be modified to cater for changing requirements or preferences.

The specification of spatial rules and grammars leads naturally to the generation and exploration of possible designs; spatial elements emerging under a part relation is highly enticing to design search [8, 9]. However, the concept of search is more fundamental to design than its generational form alone might imply. In fact, any mutation of an object into another, or parts thereof, can constitute an action of search. As such, a rule can be considered to specify a particular compound operation or mutation, that is, a composition of operations and/or transformations that is recognized as a new, single, operation and applied as such. Similarly, the creation of a grammar is merely a tool that allows a structuring of a collection of rules or operations that has proven its applicability to the creation of a certain set (or language) of designs.

Sortal Behaviors

The simplest specification of a part relationship corresponds to the subset relationship in mathematical sets. Such a part relationship applies to points and labels, e.g., a point is part of another point only if they are identical, and a label is a part of a collection of labels only if it is identical to one of the labels in the collection. Here, *sortal* operations of addition, subtraction, and product correspond to set union, difference, and intersection, respectively. In other words, if x and y denote two data forms of a *sort* of points (or labels), and X and Y denote the corresponding sets of data elements, i.e., sets of points (or labels), then (x: X specifies X as a representation of x)

$$\begin{aligned} x : X \wedge y : Y \Rightarrow x''y \Leftrightarrow X &\subseteq Y \\ x + y : X &\bigcup Y \\ x - y : X &/ Y \\ x \cdot y : X &\bigcap Y \end{aligned}$$

An alternative behavior applies to weights (e.g., line thicknesses or surface tones) as is apparent from drawings on paper—a single line drawn multiple times, each time with a different thickness, appears as if it were drawn once with the largest thickness, even though it assumes the same line with other thicknesses (see also

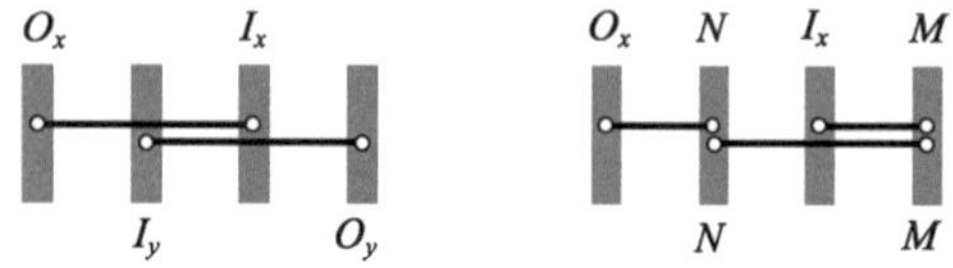

Fig. 2 The specification of the boundary collections I_x, O_x, I_y, O_y, M and N, given two data forms of intervals x (*above*) and y (*below*)

[11]). When using numeric values to represent weights, the part relation on weights corresponds to the less-than-or-equal relation on numeric values;

$$\begin{aligned} x:\{m\} \wedge y:\{n\} \Rightarrow x'' y \Leftrightarrow m'' n & \\ x+y:\{\max(m,n)\} & \\ x-y:\{\}\mathit{if}\ m'' n,\text{else}\{m\} & \\ x\cdot y:\{\min(m,n)\} & \end{aligned}$$

Thus, weights combine into a single weight, with its value as the least upper bound of the respective individual weights, i.e., their maximum value. Similarly, the common value (intersection) of a collection of weights is the greatest lower bound of the individual weights, i.e., their minimum value. The result of subtracting one weight from another depends on their relative values and is either the first weight, if it is greater that the second weight, or zero (i.e., no weight).

Another kind of part relationship corresponds to interval behavior. Consider, for example, the specification of a part relationship on line segments. A line segment may be considered as an interval on an infinite line (or carrier); in general, one-dimensional quantities, such as time, can be treated as intervals. An interval is a part of another interval if it is embedded in the latter; intervals on the same carrier that are adjacent or overlap combine into a single interval. Specifically, a behavior for intervals can be expressed in terms of the behavior of the boundaries of intervals. Let $B[x]$ denote the boundary of a data form x of intervals and, given two data forms x and y let I_x denote the collection of boundaries of x that lie within y, O_x denote the collection of boundaries of x that lie outside of y, M the collection of boundaries of both x and y where the respective intervals lie on the same side of the boundary, and N the collection of boundaries of both x and y where the respective intervals lie on opposite sides of the boundary (Fig. 2) [12]. Then,

$$\begin{aligned} x:B[x] \wedge \mathrm{y}:B[y] \Rightarrow x'' y \Leftrightarrow I_x = 0 \wedge O_y = 0 \wedge N = 0 & \\ x+y:B[x+y] = O_x + O_y + M & \\ x-y:\mathrm{B}[\mathrm{x}-\mathrm{y}] = O_x + I_y + N & \\ x\cdot y:B[\mathrm{x}\cdot y] = I_x + I_y + M & \end{aligned}$$

This behavior applies to indefinite intervals too, providing that there is an appropriate representation of both (infinite) ends of its carrier. Likewise, behaviors can be specified for area intervals (plane segments) and volume intervals (polyhedral segments). The equations above still apply though the construction of I_x, O_x, I_y, O_y, M, and N is more complex [12].

Exemplar Grammar Systems

A uniform characterization for a variety of grammar systems is given in [1]. Krishnamurti and Stouffs [13] survey a variety of spatial grammar formalisms from an implementation standpoint. Here, we consider the specification of some of these examples using *sorts*.

Structure Grammars

Structure grammar is an example of a set grammar. "A structure is a symbolic representation of parts and their relationships in a configuration" [3]. A *structure* is represented as a set of pairs, each consisting of a symbol, e.g., a spatial icon, and a transformation. The resulting algebra corresponds to the Cartesian product of the respective algebras for the set of symbols and the group of transformations. Both symbols and transformations define *sorts* with discrete behavior, i.e., respective sets match under the subset relationship. These combine into a composite *sort* under the attribute relationship; each symbol in a set may have one or more transformations assigned as an attribute.

```
sort symbols : [ImageUrl];
sort transformations : [Transformation];
sort structures : symbols ^ transformations;
```

The *sort* symbols is specified to use the *sortal* template ImageUrl, a variant on the template Label that allows the label to be treated as a URL pointing towards an image that can be downloaded and displayed.

Tartan Worlds

Tartan Worlds [14] is a spatial grammar formalism that bestrides string and set grammars. We consider a simplified string grammar version of the *Tartan Worlds*: each symbol in a string corresponds to a geometrical entity represented as a graphical icon and located on a grid. A rule in these *simplified Tartan Worlds* [13] consists of one symbol on the *lhs* and symbols on the *rhs* given in their spatial relation. An equivalent *sortal* grammar may be defined over a *sort* composed over a grid of a *sort* of graphical icons. On a fixed-sized grid, the behavior of the composite *sort* breaks down into the behavior of the *sort* of graphical icons, e.g., ordinal or discrete, over each grid cell. The matching relation is defined in the same way.

```
sort icons : [ImageUrl];
sort tartan_worlds : icons {30, 20};
```

Again, the *sort* icons is specified to use the *sortal* template ImageUrl, The *sort* tartan_worlds is defined as a composition of the *sort* icons over a fixed-size grid, similar to a two-dimensional array, of 30 by 20.

Augmented Shape Grammars

A *shape* [1] is defined as a finite arrangement of spatial elements from among points, lines, planes, or volumes, of limited but non-zero measure. A shape is a part of another shape if it is embedded in the other shape as a smaller or equal element; shapes adhere to the maximal element representation [15, 16]. Shapes of the same dimensionality belong to the same algebra; these define a *sort*. A shape consisting of more than one type of spatial elements belongs to the algebra given by the Cartesian product of the algebras of its spatial element types. The respective *sorts* combine under the operation of sum, as a disjunctive composition.

A shape can be augmented by distinguishing spatial elements, e.g., by labeling, weighting, or coloring these elements. Augmented shapes also specify an algebra as a Cartesian product of the respective shape algebra and the algebra of the distinguishing attributes. However, the resulting behavior can better be expressed with a *sort* that is a subordinate composition of the respective *sorts*, i.e., combined under the attribute operator. A *sort* of labels may adhere to a discrete behavior, a *sort* of weights to an ordinal behavior; a weight matches another weight if it has a smaller or equal value.

Most shape grammars only allow for line segments and labeled points:

```
  sort line_segments : [LineSegment];
  sort labeled_points : (points : [Point]) ^ (labels :
[Label]);
  sort shapes : line_segments + labeled_points;
```

Sortal Rules

When considering a simple *sortal* grammar, the grammar rules can all be specified within the same *sort* as defined for the grammar formalism. In the case of more complex *sortal* grammars, or when the grammar formalism may change or develop over time, it may be worthwhile to consider grammar rules that are specified within a different *sort*, for example, a simpler *sort* or a previously adopted *sort*, without having to rewrite these to the *sortal* formalism currently adopted. *Sortal* grammar formalisms support this through the subsumption relationship over *sorts*. This subsumption relationship underlies the ability to compare *sortal* representations. and assess data loss when exchanging data from one *sort* to another. When a representation subsumes another, the entities represented by the latter can also be represented by the former representation, without any data loss.

Under the disjunctive operation of sum, any entity of the resulting *sort* is necessarily an entity of one of the constituent sorts. *Sortal* disjunction consequently defines a subsumption relationship on *sorts* (denoted '$\leq$'), as follows:

$$a \leq b \Leftrightarrow a + b = b;$$

a disjunctive *sort* subsumes each constituent *sort*.

Most logic-based formalisms link subsumption directly to information specificity, that is, a structure is subsumed by another, if this structure contains strictly more information than the other. The subsumption relationship on *sorts* can also be considered in terms of information specificity, however, there is a distinction to be drawn in the way in which subsumption is treated in *sorts* and in first-order logic based representational formalisms. First-order logic formalisms generally consider a relation of inclusion (hyponymy relation), commonly denoted as an is-a relationship. *Sorts*, on the other hand, consider a part-of relationship (meronymy relation).

Two simple examples illustrate this distinction. Consider a disjunction of a *sort* of points and a *sort* of line segments; this allows for the representation of both points and line segments. We can say that the *sort* of points forms part of the *sort* of points and line segments—note the part-of relationship. In first-order logic, this corresponds to the union of points and line segments. We can say that both are bounded geometrical entities of zero or one dimensions—note the is-a relationship.

This distinction becomes even more important when we consider an extension of *sortal* subsumption to the attribute operator. Consider a *sort* cost_types as a composition under the attribute relationship of a *sort* types with template Label and a sort costs with template Weight:

```
  sort  cost_types  :  (types  :  [Label])  ^  (costs  :
[Weight])
```

For example, these cost values may be specified per unit length or surface area for building components. If we lessen the conjunctive character of the attribute operator by making the cost attribute entity optional, then, we can consider a type label to be a cost type without an associated cost value or, preferably, a type label to be part of a cost type, that is, the sort of types is part of the sort of cost types. Vice versa, the sort of cost types subsumes the sort of types or, in general:

$$a \leq a \wedge b$$

In logic formalisms, a relational construct is used to represent such associations. For example, in description logic [17], roles are defined as binary relationships between concepts. Consider a concept Label and a concept Color; the concept of colored labels can then be represented as Label ∩ ∃ `hasAttribute.Color`, denoting those labels that have an attribute that is a color. Here, ∩ denotes intersection and ∃R.C denotes full existential quantification with respect to role R and concept C. It follows then that `Label` ∩ ∃ `hasAttribute.Color` ⊆ `Label`; that is, the concept of labels subsumes the concept of colored labels—this is quite the reverse of how it is considered in *sorts*.

As such, a shape rule specified for a *sort* of line segments and labeled points remains applicable if we extend the formalism to include plane segments or even volumes (if all considered in three dimensions). Similarly, the shape rule would still apply if we adapt the formalism to consider colored labels as attributes to the points, or line segments for that matter.

Another important distinction is that first order logic-based representations generally make for an open world assumption—that is, nothing is excluded unless it is done so explicitly. For example, shapes may have a color assigned. When looking for a yellow square, logically, every square is considered a potential solution—unless, it has an explicitly specified color, or it is otherwise known not to have the yellow color. The fact that a color is not specified does not exclude an object from potentially being yellow. As such, logic-based representations are automatically considered to be incomplete. *Sorts*, on the other hand, hold to a closed world assumption. That is, we work with just the data we have. A shape has a color only if one is explicitly assigned: when looking for a yellow square, any square will not do; it has to have the yellow color assigned. This restriction is used to constrain the application of grammar rules, as in the use of labeled points to constrain the application of shape grammar rules. Another way of looking at this distinction between the open or closed world assumptions is to consider their applicability to knowledge representation. To reiterate, logic-based representations essentially represent knowledge; *sorts*, on the other hand, are intended to represent data—any reasoning is based purely on present (or emergent) data.

Urban Design Grammars

Beirão, Duarte and Stouffs [18] present components of an urban design grammar inferred from an extension plan for the city of Praia in Cabo Verde (Fig. 3). The development of the urban grammar forms part of a large research project called City Induction aiming at integrating an urban program formulation model [19], a design generation model [20] and an evaluation model [21] in an 'urban design tool'. The central idea to the project is to read data from the site context on a GIS platform, generate program descriptions according to the context conditions, and from that program generate alternative design solutions guided by evaluation processes in order to obtain satisfactory design solutions. The generation part considers urban grammars in an extension of the discursive grammar schema developed by Duarte [22] and adapted for urban design. The generation of urban designs is led by the selection and application of urban patterns (denoted Urban Induction Patterns), each formalized as a discursive grammar. The application of an urban pattern—or discursive grammar—involves the application of urban design rules codified in a spatial grammar according to the requirements of the urban program codified in a description grammar [23]. The spatial grammar may manifest itself as an augmented shape grammar, allowing for various attribute data to be associated with the graphical elements, but may also include a raster-based string or set grammar, in order to allow rule-based operations on both vectorized and rasterized GIS data.

Acknowledgments The author wishes to thank Ramesh Krishnamurti for his contributions to the *sortal* research, and José Beirão and José Duarte for their collaboration on urban design grammars.

References

1. Stiny G (1980) Introduction to shape and shape grammars. Environ Plan B: Plan Des 7:343–351
2. Stiny G (1981) A note on the description of designs. Environ Plan B: Plan Des 8:257–267
3. Carlson C, McKelvey R, Woodbury RF (1991) An introduction to structure and structure grammars. Environ Plan B: Plan Des 18:417–426
4. Duarte JP, Correia R (2006) Implementing a description grammar: generating housing programs online. Constr Innov: Inf Process Manag 6(4):203–216
5. Duarte JP (2005) A discursive grammar for customizing mass housing: the case of Siza's houses at Malagueira. Autom Constr 14:265–275
6. Stouffs R, Krishnamurti R (2001) Sortal grammars as a framework for exploring grammar formalisms. In: Burry M, Datta S, Dawson A et al. (eds) Mathematics and design 2001. The School of Architecture & Building, Deakin University, Geelong, pp 261–269
7. Stouffs R (2008) Constructing design representations using a sortal approach. Adv Eng Informatics 22(1):71–89
8. Mitchell WJ (1993) A computational view of design creativity. In: Gero JS, Mahel ML (eds) Modeling creativity and knowledge-based creative design. Erlbaum, Hillsdale
9. Stiny G (1993) Emergence and continuity in shape grammars. In: Flemming U, Van Wyk S (eds) CAAD futures 1993. North-Holland, Amsterdam, pp 37–54

10. Gips J, Stiny G (1980) Production systems and grammars: a uniform characterization. Environ Plan B: Plan Des 7: 399–408
11. Stiny G (1992) Weights. Environ Plan B: Plan Des 19: 413–430
12. Krishnamurti R, Stouffs R (2004) The boundary of a shape and its classification. J Design Res 4(1)
13. Krishnamurti R, Stouffs R (1993) Spatial grammars: motivation, comparison and new results. In: Flemming U, Van Wyk S (eds) CAAD Futures '93. North-Holland, Amsterdam, pp 57–74
14. Woodbury RF, Radford AD, Taplin PN et al. (1992) Tartan worlds: a generative symbol grammar system. In: Noble D, Kensek K (eds) ACADIA '92
15. Krishnamurti R (1992) The maximal representation of a shape. Environ Plan B: Plan Des 19:585–603
16. Stouffs R (1994) The algebra of shapes. Ph.D. dissertation. Department of Architecture, Carnegie Mellon University, Pittsburgh
17. Baader F, Calvanese D, McGuinness D et al (2003) The description logic handbook: theory, implementation and applications. Cambridge University, Cambridge
18. Beirão J, Duarte J, Stouffs R (2009) An urban grammar for Praia: towards generic shape grammars for urban design. In: Computation: The new realm of architectural design. Istanbul Technical University, Istanbul, pp 575–584.
19. Montenegro NC, Duarte JP (2008) Towards a computational description of urban patterns. In: Muylle M (ed) Architecture 'in computro', integrating methods and techniques. eCAADe and Artesis Hogeschool Antwerpen, Antwerp, pp 239–248
20. Beirão J, Duarte J, Stouffs R (2008) Structuring a generative model for urban design: linking GIS to shape grammars. In: Muylle M (ed) Architecture 'in computro', integrating methods and techniques. eCAADe and Artesis Hogeschool Antwerpen, Antwerp, pp 929–938
21. Gil J, Duarte JP (2008) Towards an urban design evaluation framework. In: Muylle M (ed) Architecture 'in computro', integrating methods and techniques. eCAADe and Artesis Hogeschool Antwerpen, Antwerp, pp 257–264
22. Duarte JP (2005) A discursive grammar for customizing mass housing: the case of Siza's houses at Malagueira. Autom Constr 14(2):265–275
23. Stiny G (1981) A note on the description of designs. Environ Plan B: Plan Des 8(3):257–267

Raising the *i*-Factor: Bridging Parametric Shape and Parametric Design

Ramesh Krishnamurti

The i-factor

I research software for computer-aided design. In this connection, whenever I talk to my students about design software I often use the *i*-device as the modern metaphor for inquiry, ideation, instruction, interaction, interchange, interface, and not least, information. It is the harbinger of change and quite possibly the symbol of present-day instant gratification. The *i*-device appears to allow people to work seamlessly and simultaneously with multiple technologies towards a single collective task. Prior to its advent, each individual technology was largely employed for specialized concerns; whereas now, seemingly disparate technologies work well together with what I term high *i-factor*. There are, of course, technologies that have and may always have a low *i*-factor; and others about which we remain unsure. Design research has many that fall into this last category.

Computer-aided design software has wide appeal. Much of this appeal has to do with the fact that most modeling software deal with geometry and shape manipulation, with some offering an added ability to create design information models. Design for use, fabrication, or construction requires renderings, performance analyses in a multitude of domains, for example, lighting, energy, acoustics, fluid flow, materials and sustainability; or surface deconstruction for fabrication. They require software with specialized (that is, less) appeal. Combining the two has cost considerations—representation cost, interchange cost, process cost, integration cost and so on. Moreover, the costs would be disproportionately skewed towards higher costs designs. That is, performance software currently has a low *i*-factor if one considers a larger client (or designer) base. To work seamlessly together, to integrate multivariate requirements we would have to readdress issues pertaining to design representation, design information and/or design processes. Figure 1 is illustrative of agents and entities in a design process. It is also suggestive of the sorts of problems and models worth exploring,

R. Krishnamurti (✉)
Carnegie Mellon University, Pittsburgh, USA
e-mail: ramesh@cmu.edu

J. S. Gero (ed.), *Studying Visual and Spatial Reasoning for Design Creativity*,
DOI 10.1007/978-94-017-9297-4_5

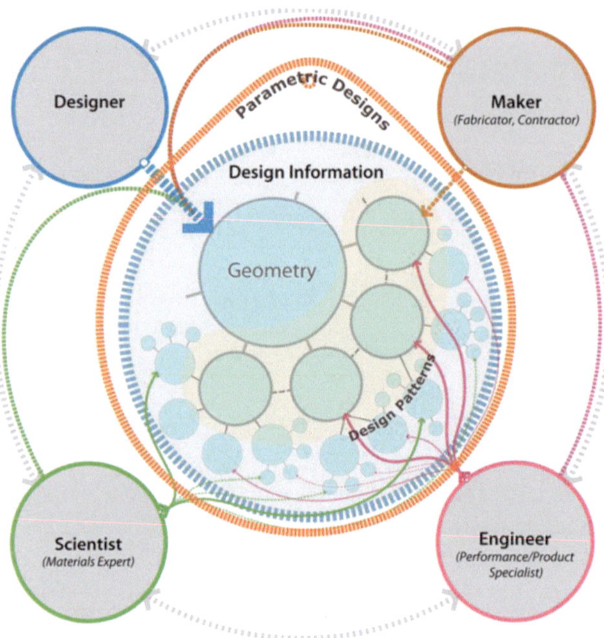

Fig. 1 Agents and entities in a design process

for instance, compositional explorations of geometry; formal and functional explorations via parametric design, design patterns and physics-based simulation; and physical manifestations through element decomposition, fabrication and/or assembly.

For the past few years I have been engaged in two distinct types of design research, one with a practical bent but requiring more *theoretical* consideration; the other based much more solidly in theory but calling out for *real* practical application. Both types of research provide fresh insights. Interestingly, one arrives at quite similar conclusions to quite distinct problems. And, not least, judging by the results/products from earlier research in these areas, these appear to have low *i*-factor. The question is: *what does it take to raise the i-factor*?

There are a number of ways in which to explore this issue. One via cognition is on how users perceive and manipulate information and tools that are already provided by design software and in finding ways of making such information accessible, consequently affording new and perhaps novel uses. Note here that I am not questioning the nature and structure of information; these are taken as given, instead the focus is positioned on how the representation of different sorts of information allow for meaningful combinations that will assist the designer's intentions. As a simple recent example, interfaces now visually and dynamically allow users to interact with and change sun paths, rapidly calculating consequences on shades and shadows, thereby enabling users to make much informed, versatile on-the-fly design decisions. Sun path technology is established; newer models of interactions afford interesting and/or novel sun-tempered applications. Commercial design software is increasingly adopting this route towards redesigning the way known information is envisioned and manipulated.

Another, but equally important consideration is at the other extreme in looking at the underlying technology itself and seeing ways in which improved information access and use can be facilitated. This approach requires an investigation of alternative ways information can be structured, while taking into account any constraints in current design software and technologies. For instance, in digital fabrication there is a growing demand on how information that is not currently provided in the design software can be incorporated and accessed by the designer. This need does not involve simply finding ways to represent information but also questioning the current design tools and suggesting novel ways to structure information forming a rationalized process.

The idea underlying much of my work is simple, namely, if different sorts of information were made easily accessible to designers, they might then use them to solve a range of other problems that their computational design technology was not originally intended for. That is, raising the *i*-factor. In each case, it necessitates a design model and (designer friendly) programmable computational environment. In the sequel I highlight some of the design software projects that my students and I have worked on and suggest in each case possible ways of raising the *i*-factor.

Parametric Sustainable Building Design

In a typical building design, evaluating performances of a design information model-based solution requires a heavy dose of human intervention and data interpretation, rendering analyses to be both costly and time consuming. Evaluating for sustainability according to a green building design standard not only requires different types of analyses but also requires evaluation of different aspects of the building. For this we used Revit®—with its advanced parametric and building modeling capabilities—as the design information model in an integrated database [1]. We used its environment to develop a prototype (Fig. 2), which is focused on how users interact with design and evaluation with respect to a green building rating standard.

The purpose of the project is to provide real-time feedback on the status of a building project with respect to sustainability according to a chosen green building rating standard. Information required for evaluating the sustainability of the project is gathered from the building information model, external databases and simulation results. The prototype goes through three steps during an evaluation: (i) check pre-requisite credits; (ii) supply additional simulation results; and (iii) evaluate credit.

An evaluation starts once pre-requisite credits have been checked. The prototype retrieves information to evaluate credits for the currently selected standard, which for the example in Fig. 2 is LEED [2]. Any added data is then supplied, for example,

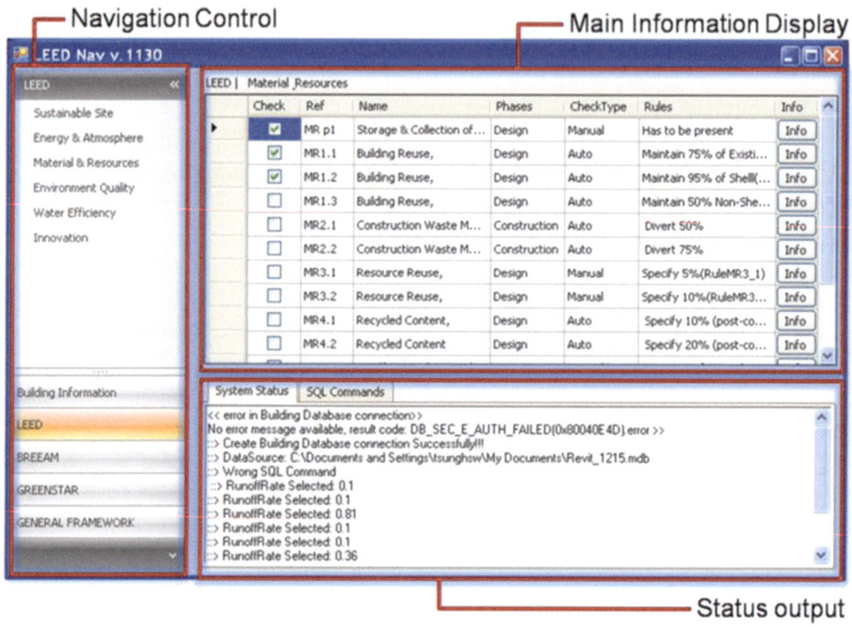

Fig. 2 Prototype to evaluate buildings with a green building rating standard

simulation results. On aggregating the required data from the sources, namely, the building information model, simulation results and rating standard, results are updated to a data table.

To validate the implementation we chose an existing LEED Silver-certified building to run an evaluation based on a Revit model and real simulation results. We tried to maximize the possible credits achievable by automating the process. Figure 3 shows results for the LEED Material Resources category. The prototype can generate a graphical representation of the evaluation results in three different formats: as a table, an image, and as an html file. These results provide users with the current status of the project.

On face value, all this appears straightforward: a design information model seemingly linked seamlessly to performance evaluation applications and a green standard. *So what's missing? Or rather, what does it take to accomplish this?* The answer lies quite literally in filling the gaps.

Filling the Gaps for Sustainable Design

A design information model acts as a data container to hold design and other relevant information. However, currently, these models contain less than sufficient data to meet most rating system requirements. Some require data external to the design information model; these have to be accommodated in a cohesive manner

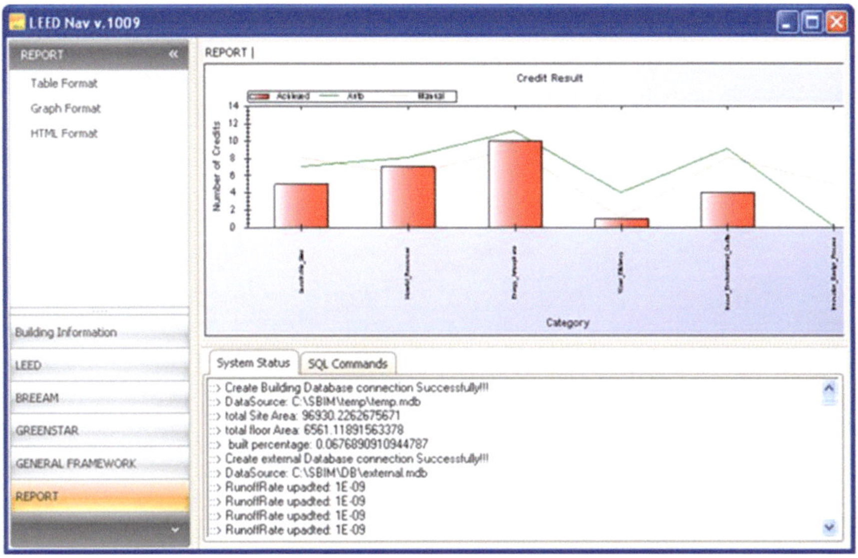

Fig. 3 LEED material and resources credit evaluation and report generation

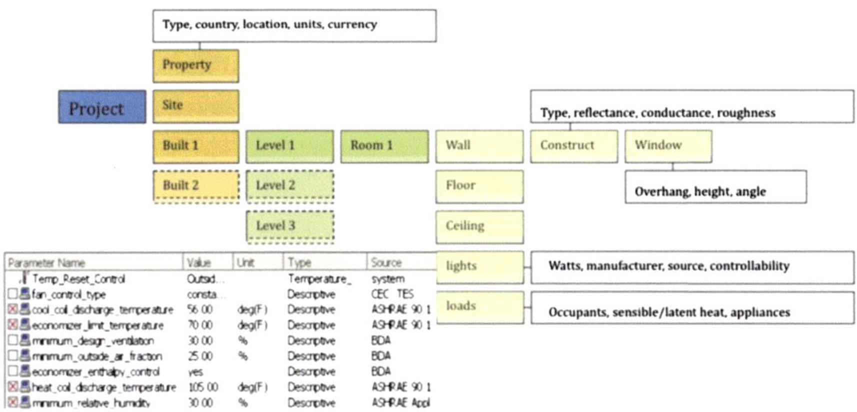

Fig. 4 Design elements for energy analysis not typically found in any design model

[3]. The data might be electronically and geographically distributed; they may need to be salvaged; and they may even need to be certified. See Fig. 4.

Our approach was to develop a framework for mapping elements and measures to evaluate each credit of a green building rating system to elements and measures in the design information model [4]. The framework also identifies information not accessible within the design model. We also developed a visualization tool to exemplify how the available information can be used to guide and create a design from a sustainability viewpoint. The functionalities of the visualization tool, shown

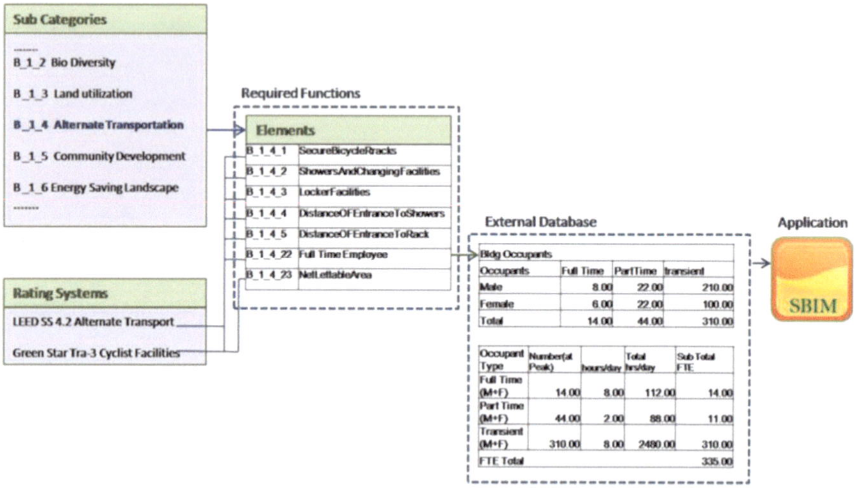

Fig. 5 Snapshot of mapping rating system requirements to elements in the model

in Fig. 6 allow us to look at a chosen design, which has been evaluated by the prototype application, and show the nature of information used in evaluating chosen credits. In the window the top-level categories of the chosen rating system is shown on the left, here LEED; the next level down shows credits belonging to the category. Credits are color-coded to reflect the level of achievement when evaluated by the Revit application: red for a fully achieved credit, white when it has not, and pink when it almost has.

If missing design elements, supporting databases and simulation results are available, credits are clearly achievable. This tool focuses on required design elements for evaluation rather than mirroring results from the prototype application. It can be used in two ways: as an extension to the prototype, or to invoke Revit directly for the current project. That is, the gap analysis tool can be used to template automated sustainability validation (Fig. 5).

In the spirit in which this paper is set out, the obvious next step is to ask how much further can one push the boundaries of integration. The next project does just that by exploring this single building design performance evaluation application at the urban scale.

Parametric Sustainable Urban Water Use

Calculating water use is straightforward. However, it can be problematical because of missing data when integrating sustainability requirements with a particular design model. Pertinent data for water use calculations include occupant numbers, fixture flow rates, fixture costs and materials. Designer usually supplies these. Other required data outwith the design information model include rainfall, plant water

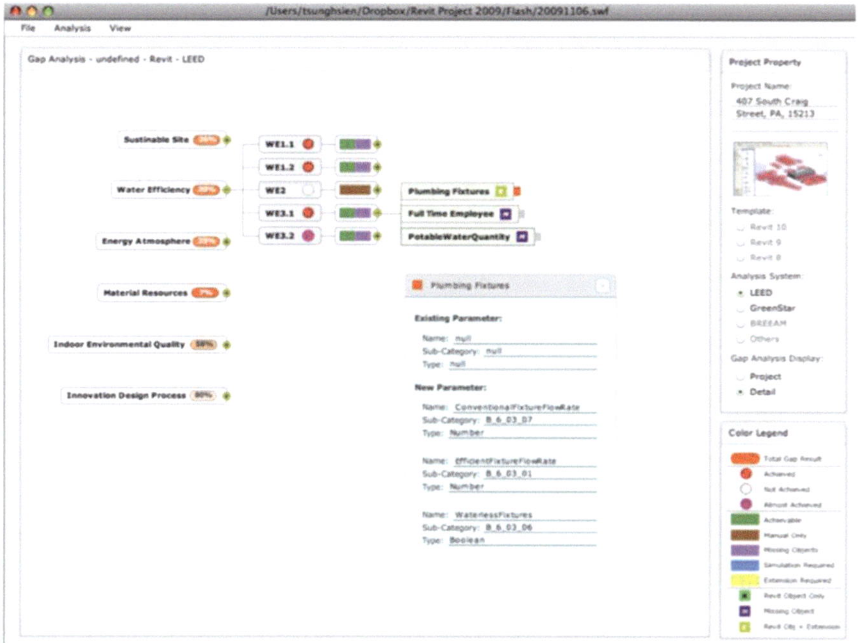

Fig. 6 Gap visualization of LEED WE3.1 illustrating extensions

use, etc. Such data do not normally to fall within the designer's purview, yet these are all factors that have to be accounted for. For this project we focused on modeling sustainable water use on an urban scale. The introduction of scale leads us to consider two approaches aimed at fulfilling LEED requirements for water use reduction (Fig. 7).

Approach 1 The first approach is based on the hypothesis that if water use can be calculated and evaluated for a single building it can be extended to multiple buildings and thus, to a larger scale. For this we modeled a single commercial building in Revit using our prototype application to retrieve relevant information, namely, building heights and occupants.

Calculations mainly follow the LEED method for water use, which is found by estimating occupant usage and fixture flow rates [5]. Occupant use corresponds to full time equivalent (FTE) occupancy of a building, based on a standard 8-hour occupancy period, resulting in a value based on an eighth of the hours per day. For transient building populations such as students, visitors, or customers, hours are estimated on a representative daily average. Water use reduction for a building then corresponds to the difference between a design case and the baseline case [5].

A *design* case is obtained by totaling the annual volume of water use per fixture type and then, subtracting rainwater or gray water reuse. Actual flow rates and flush volumes for the installed fixtures are used in the calculation. For consistency, a

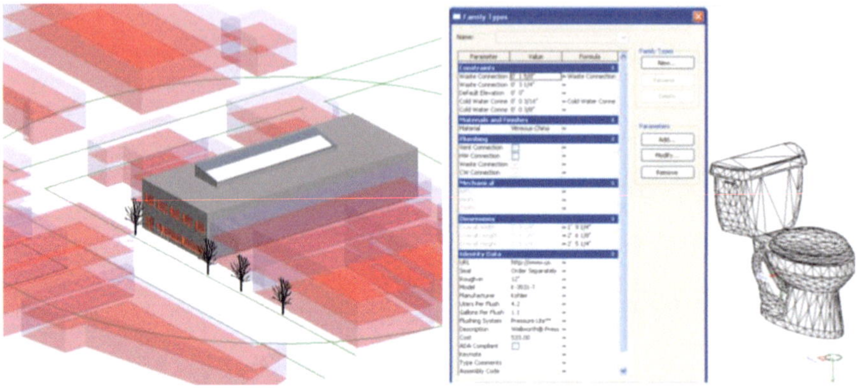

Fig. 7 (*Left*) A test case for modeling water; (*right*) fixture information

balanced one-to-one ratio of male and female is assumed. To create the *baseline* case, the design case information is used to provide the number of male and female occupants, with fixture flush and flow rate values adjusted as per EPAct default specifications [5].

Water fixtures components are stored in the Revit library and can be queried for dimensions. Other element parameters such as fixture flow rate, fixture cost, have to be filled unless automatically generated from manufacturers specifications.

To calculate water efficiency, we implemented external databases for fixtures and landscapes. Additional materials and element parameters include material porosity and fixture flow rates. Fixture costs from manufacturers are used for comparison in water use and ultimate cost savings. In the prototype there are two tabs for water efficiency. See Fig. 8. These contain the necessary tasks for evaluating water efficiency credits.

The overall workflow for water use retrieves information about the numbers of male and female occupants, which are specified in the Revit model. Differences between the baseline and design cases are then compared to determine the number of credits earned at this stage.

This approach can be extended to accumulate information on multiple buildings and aggregate total water use. This approach works only when all pertinent information is available.

Approach 2 The second method uses a two-dimensional drawing to generate a mass model augmented with water related information. In modeling water use on an urban scale, where we have information on building area and height, we employ a combination of different software. As before, only fixture flow rates and occupant use are considered in water use calculations ignoring such quantities as gray water quantity and rainwater harvesting. With many buildings in an urban area, assigning occupants to each building is difficult; in this case, we employ Green Star's method, based on net usable space, for assigning occupants [6].

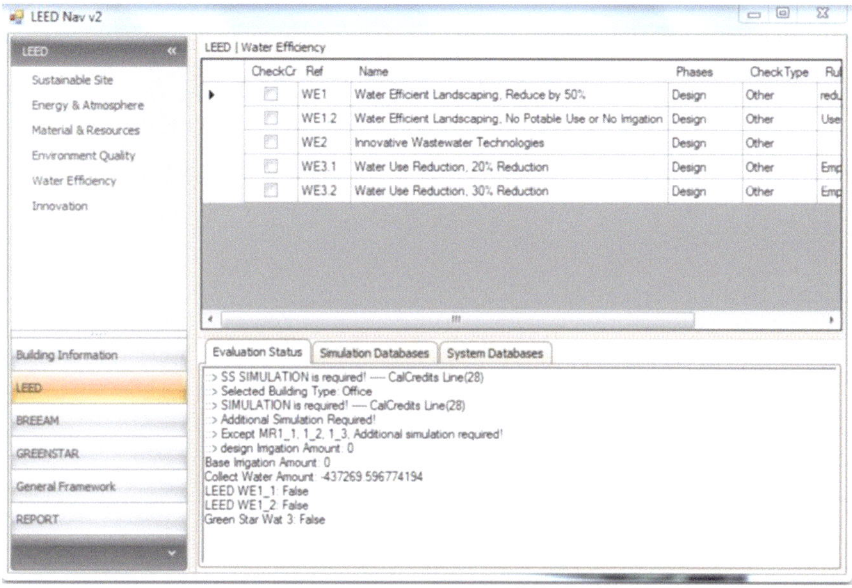

Fig. 8 Water efficiency tabs for rating credit and other water related calculations

The sample case study covers an area nearly 17,350 m^2. Of this about 11,700 m^2 covers the building footprint; the remainders are roads, pavements and parking areas, which are assumed to have an impervious ground cover. There are open spaces with potential for planning for rainwater catchments and water management [7]. The mass model (Fig. 10) is generated from a CAD drawing (Fig. 9), from which total floor areas of buildings can be calculated. We used Rhinoceros® [8] with Grasshopper™ [9] for the mass model.

This approach offers a way to easily create a parametric model with facility to calculate quantities and specify parameters for water use calculation on a larger scale with greater flexibility than Revit currently affords. As both Rhinoceros and Revit are built on top of the .NET framework [10], communicating between software is straightforward. The Grasshopper definitions in Fig. 11 shows the connection between model geometry and the fixtures database to generate an urban water use model.

Remarks

Both approaches are parametric allowing variations in fixture number and type, ratio of male to female occupants, and in allocating different design cases to different parts of the urban area in order to model various water use scenarios. The combination of two commercial software, external databases, and sustainability requirements illustrate how information can be gathered and processed on a larger scale. Despite

Fig. 9 CAD drawing of the test urban area

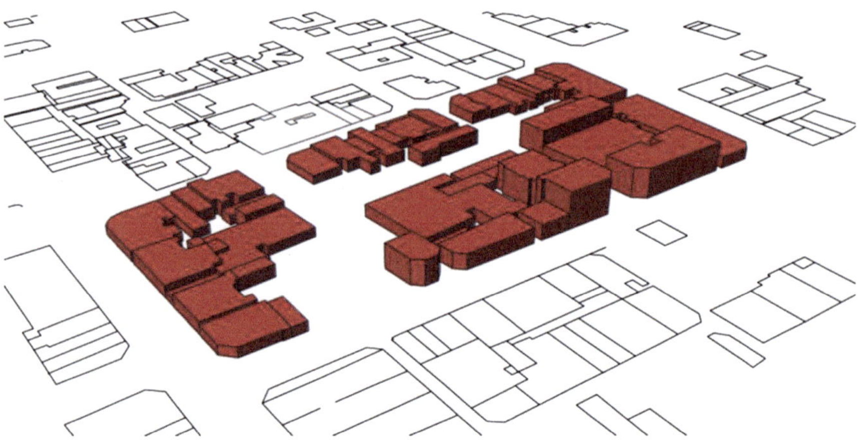

Fig. 10 Mass model of the urban area

differences in data structures and manipulations, it is possible to raise the *i*-factor by leveraging a programming environment to both pre-certify a single building for sustainability and on a larger scale, to project effects on environmental resources.

Parametric Surface Tessellation

The next project is inspired by a growing trend in contemporary architectural practice of exploiting freeform surfaces to design and model intricate geometries for projects which otherwise would be impossible to realize. In doing so, designers have liberally borrowed digital fabrication techniques developed in the automobile and aerospace industries [11–13].

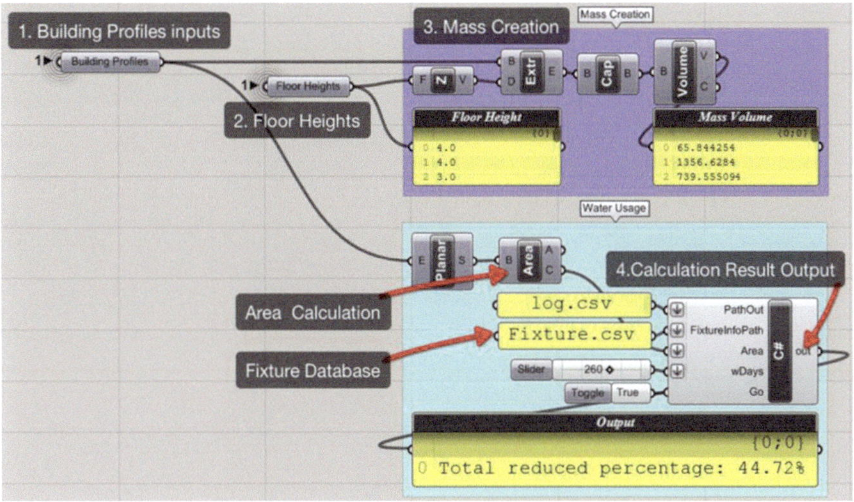

Fig. 11 Grasshopper definition file for urban water use model generation

To manifest a freeform surface, a discrete mesh model is utilized. Transforming a freeform surface into an appropriate mesh is computationally intensive; generally, it is not an easy for designers with no formal geometry training. To design, model, and, subsequently, fabricate intriguing, sometimes intricate, freeform shapes, we look at surface tessellation as an extension of meshing with the added consideration of incorporating constructible building components. There are close relationships and analogies between the elements of a mesh and the components of a freeform design.

Figure 12 shows two designs by Norman Foster and Partners: the Elephant House canopy and City Hall in London. Both demonstrate the value of the parametric approach to architectural applications. In the canopy design, the base geometry is a torus constructed by revolving a circle about an axis. The revolution gives the surface a good discretizing feature—namely, the planar quadrilateral patch [13] that can be derived directly from the principal curvature lines. This makes fabrication manageable, even for such type of doubly curved dome-like surfaces.

The City Hall project demonstrates how parametric schemes used on a conical surface development can be realized for flat panel fabrications. The initial idea for the City Hall was a pebble-like form, which was later approximated by a collection of partial conical strips, which as a member of the family of cylindrical surfaces can be easily decomposed into planar quadrilateral faces.

The two designs clearly highlight the advantages of employing constructive geometry principles in the whole process from design to fabrication. The constructive geometry principle captures the underlying form of the target surface and in turn makes feasible its ultimate manifestation.

Fig. 12 Elephant House Canopy [15]; London City Hall [16]

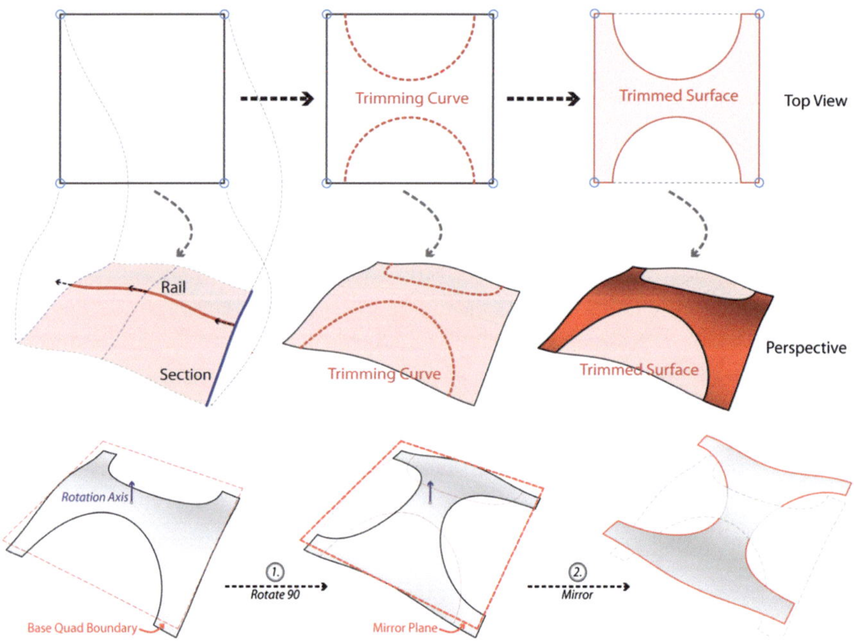

Fig. 13 Creating a trimmed surface for pattern ED_03

Interwoven Patterns

In the first instance of surface tessellations I consider interwoven patterns. In principle given a quad(rilateral)-mesh we can construct patterns (panels) by trim operations and spatial transformations.

Consider Fig. 13. The top row illustrates in top view the steps for creating a trimmed pattern. The middle row shows the corresponding surface manipulation in three-dimensional space. The bottom row illustrates transformation rules—in this

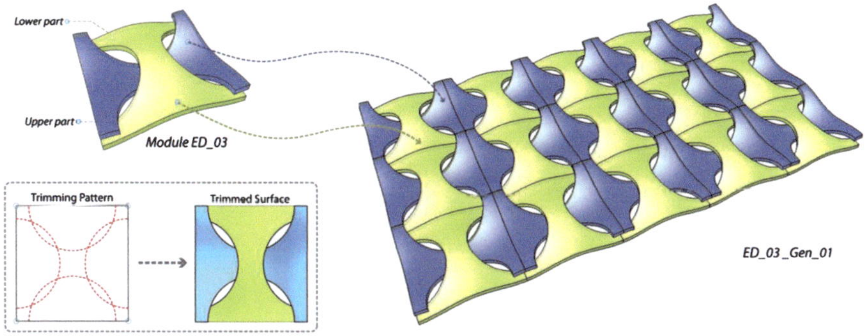

Fig. 14 The interweave pattern ED_03

case, rotation about a central axis and mirror operation via the planar quad boundary—being applied to generate the second half of this module.

Figure 14 shows the resulting interwoven pattern ED_03. The pattern is inspired by the Erwin Hauer's continuous screen, *Design 03* (circa 1952), for the Church in Leising, Vienna, Austria [17]. On the left is shown the interweave module consisting of two parts—an upper and lower module component colored in shades of green and blue respectively. The thickness of the component is derived from an offset operation in a direction normal to the base quadrilateral face.

Fig. 15 illustrates two other Hauer inspired patterns ED_04 and ED_05 again based on simple trim and spatial transformation rules. Notice that the two patterns employ the same trim patterns—two ellipses intersecting orthogonally, they generate distinct results by variation of the corresponding location to the base boundary.

The next logical step would be to consider trim rules that ensure that the basic module is self-continuous. Instead of treating a module with two separate parts, joined only at external edges with adjacent modules, this type of pattern joins parts internally. This characteristic creates a more intricate continuous movement from local module to entire modular propagation. Figure 16 illustrates such a trimming operation applied on a target surface with customized trim curves at four corners, which can be exploited to create a self-interlocking pattern.

Interwoven patterns show the application of procedure-based (or rule-based) approaches to design exploration. These patterns generalize to a parametric framework, identifying generative rules that can further manipulated, thus providing strategies for designers to develop their own parametric modeling toolkits.

Surfaces with Irregular Boundary Conditions

Real freeform designs tend not to be regular. Rather surfaces have openings introduced to meet specific design intentions, for example, lighting, viewing or circulation, etc.

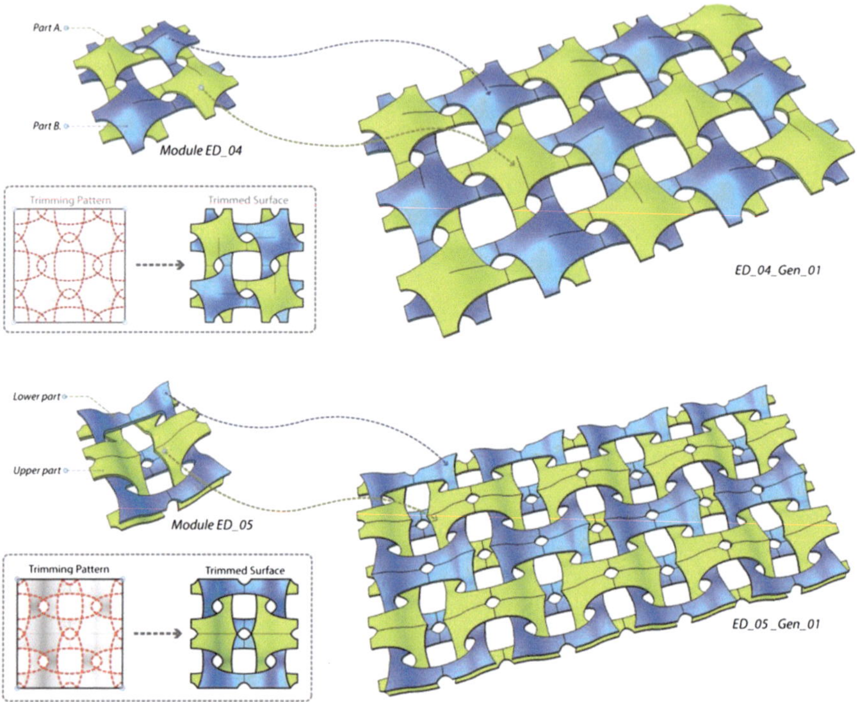

Fig. 15 Interweave patterns ED_04 and ED_05

Figure 17 shows the west elevation of a surface trimmed for skylight, natural view and entrance. Problems immediately occur when new boundaries are introduced to an originally untrimmed surface. The trimmed edges, for example, cut through the uniform shapes of certain quad dominant surface panels. Irregular panels surrounding the trim edges are generated and these, in turn, affect the overall aesthetic appearance of the surface manifestation as well as final fabrication.

To address such issues we need to explore how the global boundary conditions, which primarily determine the final freeform surface, can be used to affect or tune the surface tessellation process and be directed towards a more balanced solution. For example, directions of panelizing could be modified (or better instructed) to avoid, or reduce, the irregular panels as the boundary conditions evolve. This is ongoing work. Boundaries define the ultimate appearance; given that panelization can take all boundaries into account parametrically and algorithmically, our contention is that a coherent surface tessellation can always be achieved, and amenable to the application of design patterns [18] (Fig. 18).

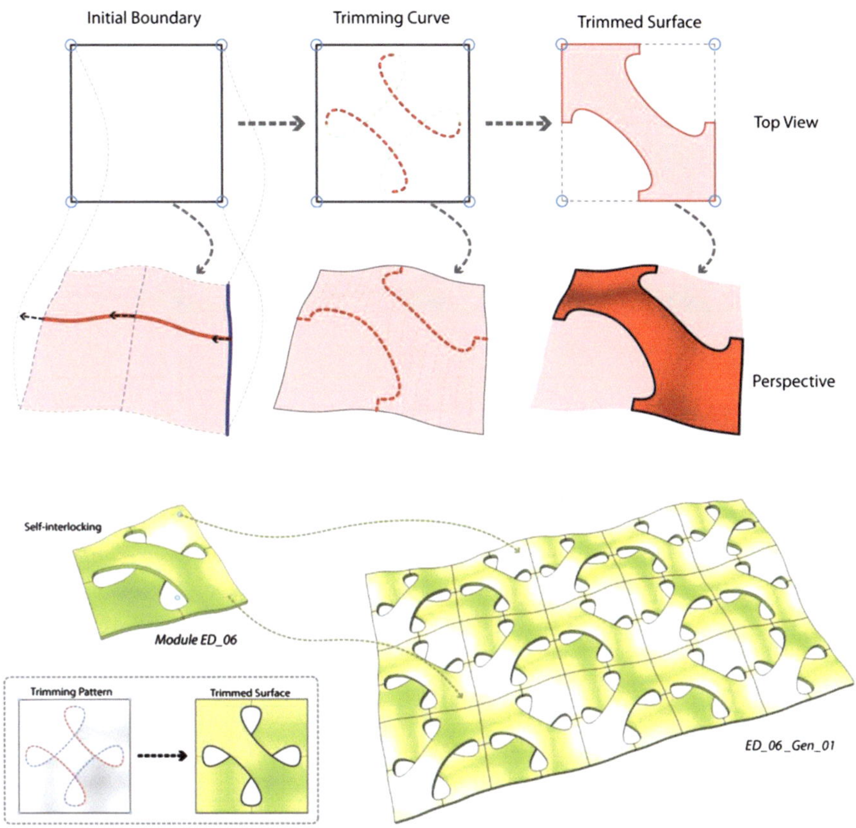

Fig. 16 Constructing the self-interlocking interwoven pattern ED_06

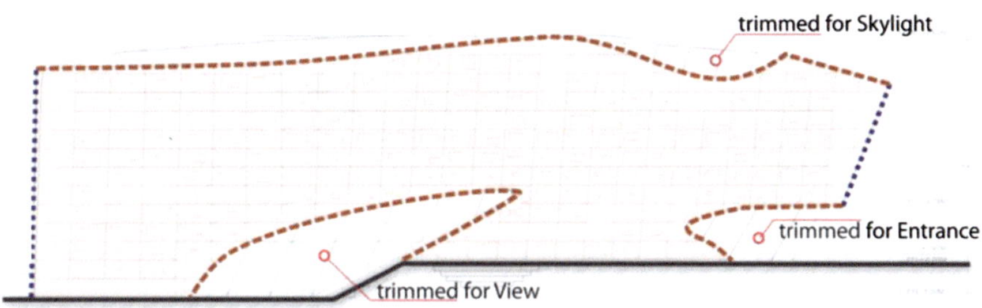

Fig. 17 Trimmed surface boundaries inspired by a Zaha Hadid design [19]

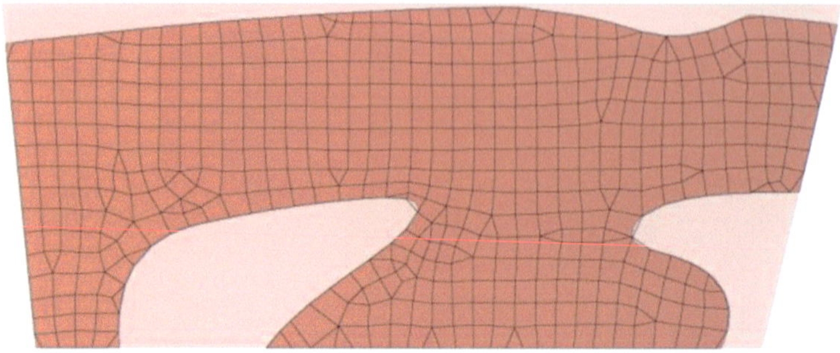

Fig. 18 A bubble meshed [20] quad-dominant tessellation

Remarks

Firstly, it should be noted that there is distinction to be drawn between this and the first project. Parametric sustainable design relies on an established technology, namely a CAD system, which by its very nature is *i*-amenable. On the other hand, parametric surface tessellation has its application in digital fabrication, which is less established, its techniques and methods are, at best, ad hoc [11, 12, 21]. Raising the *i*-factor is tantamount to providing designers with easier ways to find alternative fabrication strategies. At the very least, this is a two-stage development. From an information processing perspective, freeform surface tessellation has to become a rationalized process; only then can we consider ways of leveraging information inherent in the process.

Parametric Shape Grammars

The last project deals with shape grammars, a largely theoretical construct geared towards analyses and idea formation [22]. Shapes are created by the application of shape rules, each made up of a left and right part. Under rule application, the left part of the rule 'found' in the shape under some transformation (including parameter assignment) is replaced by the right part of the rule under the same transformation to produce a new shape. Shape grammars have been widely applied for analyzing designs [23] (Fig. 19).

Although the basic formalism of a shape grammar has remained largely the same over time, there have been changes both in definition and development. Factors that have influenced these changes relate to the scope of permissible shape elements and possible augmentations. The first formal definition, *SG1971,* was given in the seminal article by Stiny and Gips [24] More definitions have since appeared in the literature, each reflecting either the understanding at a particular time, or reflecting a specific research flavor. Definitions fall into two stages: marker-driven and subshape-driven. Definitions *SG1971*, *SG1974*, *SG1975* and

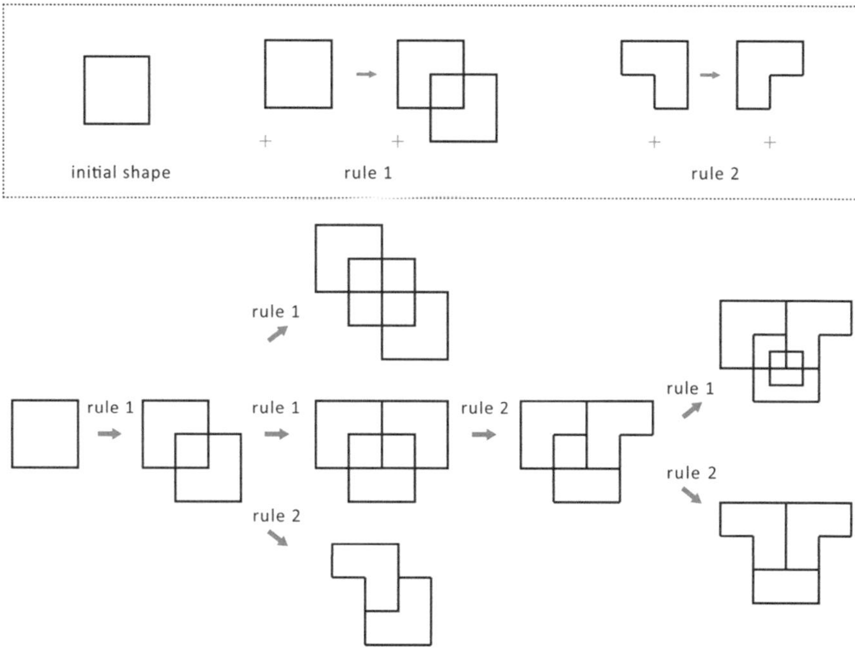

Fig. 19 An example of a shape grammar

SG1977 are marker-driven [24–27]. Definitions *SG1980, SG1991, SG1992* and *SG2006* are subshape driven [22, 23, 28, 29]. Chronologically, the definitions are backwards compatible. That is *SG1971*<<[*SG1974*, *SG1975*]<<*SG1977*<<*SG1980*<<*SG1991*<<*SG1992*<<*SG2006*, where the right side of << is more general than the left side.

Markers play a pivotal role in determining the shape rule applicability and their corresponding transformation. Markers can ensure that rule application can be directly computed. By comparison, there are harder computational issues involved with subshape-driven grammars, particularly, parametric subshape recognition and indeterminacy.

The evolutionary development exhibits a trend from 'rigid' to 'soft'. 'Rigid' here means that the shape grammars were defined to be closer to phrase structure grammars [30]. Such shape grammars are more machine-bound in the sense that they are easier to compute, but harder to use to create novel designs. A recent series of notable shape grammar implementations fall within the rigid category of shape grammars [31–34].

On the other hand, the 'softer' development is more human-centered, showing more concern and consideration on how to use shape grammars to generate novel designs. This explains, in part, the importance of subshape-driven grammars, concepts of indeterminacy and shape emergence, and the support for ambiguity in shape grammar research. Humans have little trouble handling such concepts. Moreover, human designers actually benefit from them. However, these concepts are

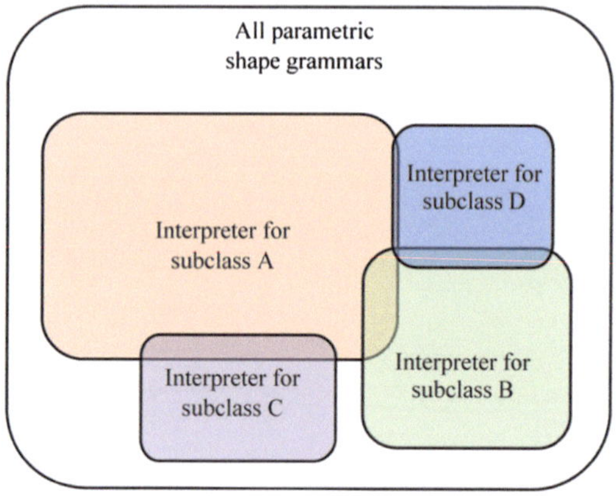

Fig. 20 A paradigm for practical "general" parametric interpreters

problematical when considering computer implementation, especially when directed at parametric shapes defined by open terms [35], though there have been notable attempts [36]. Computer implementation is vital to both research and application. The central difficulty lies in parametric subshape recognition.

In this regard, a parametric shape may have an indeterminate number, k, of open terms, that is, points with no fixed coordinates, matched against n possible points. Then, choices are combinatorial; as k approaches $n/2$, time complexity for subshape recognition is a super-polynomial.

Shape grammars fall into two distinct categories [37]. The first handles special shapes; the second is more general with potentially super-polynomial time complexity, which is only practical for shapes of smaller sizes. The implication for practice is that the best we can achieve is implementing shape grammar interpreters, each capable of handling a subset. This leads to a paradigm for practical, 'general' parametric shape grammar interpreters. See Fig. 20.

We make the assumption that interpreters for shape grammars belonging to different subclasses will collectively cover most parametric shape grammars. The classification can be considered "better" when the number of subclasses is smaller, and when, simultaneously and collectively, the scope covered is larger. Possible ways of classifying shape grammars needs further research.

We consider categories of shape grammars whose implementation is tractable. Shape grammars, which capture certain building styles, generally fall into this category. These are normally parametric shape grammars, in which rule application does not depend on emergent shapes. Markers drive rule application, and configurations are rectangular or approximated as such. Parameterization is typically limited to the height, width or room ratios. Shape rules typically relate to adding a room, subdividing a room, or refinements such as adding windows, doors, etc.

Apart from such internal characteristics, there are other factors that influence computational tractability, that is, in how shape grammars are designed and described. Normally, a shape grammar is designed to simply and succinctly describe an underlying building style, with little consideration on how it can be implemented.

As a result, in order to translate this into programming code, shape rules have to be quantitatively specified with sufficient precision to disallow the generation of ill-dimensioned configurations.

In practice there may be several ways to describing a particular shape rule; it is possible that one way is easier to compute, and another, computationally intractable. As a result, it is desirable to design an application program interface as the framework to support the design of shape grammars; then, such shape grammars are guaranteed to be computationally tractable. Such a framework requires an underlying data structure, and basic manipulation algorithms. Moreover, for ease of code translation, a meta-language built on top of the basic manipulation algorithms should be developed. As grammars in different classes typically have differing underlying structures, the appropriate underlying data structure for the framework is different. Ideally, the interpreter for a subclass of shape grammars can be supported on a single framework. Consequently, the framework for parametric shape grammars comprises a series of sub-frameworks, one for each subclass of shape grammars. Overall, as the framework is capable of ensuring tractability, we term such shape grammars as computation-friendly.

We have developed three sub-frameworks each specifying a way of implementing a subclass of shape grammars: rectangular, polygonal and graph [37]. Figure 21 illustrates a screenshot of a shape grammar interpreter based on the rectangular framework. The interface shown is an adaptation of shape grammar application in another context [38].

Remarks

Implementing parametric, indeed, non-parameterized shape grammars fall into an altogether distinct category from the first two kinds of projects. Here, theory and algorithms are well established, however, practical demonstrations have proven to be exceedingly hard. Although there are implementations of grammars [39], most cannot claim to be anything beyond a toy. Part of the difficulty involves the technicalities of shape rule application, of harnessing the power of emergent shapes to the designer's advantage. Resolving such issues will impact the *i*-factor. The approach suggested is one of classifying shape grammars in terms of local data structures, each with its own set of manipulation tools. In this way we can recast shape rule application to a limited set of functions.

Discussion

The work presented is motivated in part by wanting to extend the capabilities of the design software beyond its original intent, that is, to raise its *i*-factor. We have attempted to do so not by trying to be novel. Instead, in each project we

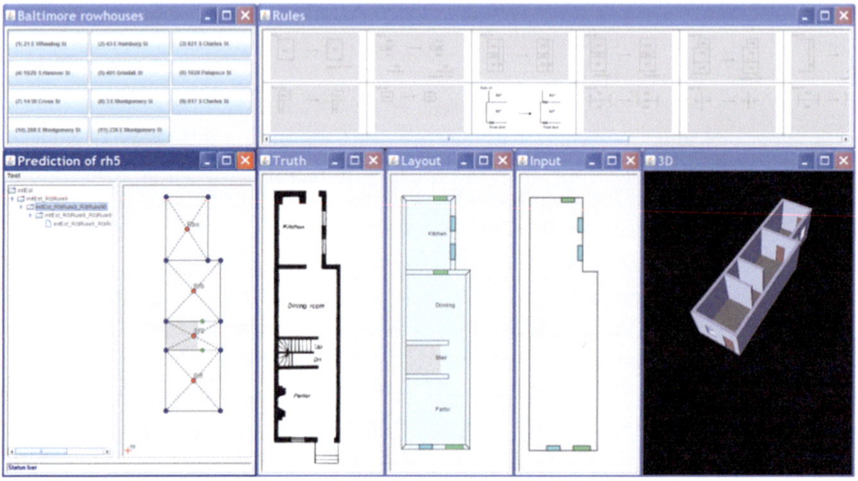

Fig. 21 Screenshot of layout determination of the Baltimore Rowhouse

have shied away from any global representation to more localized semantics by empowering the process. That is, the information in any rule is of greater significance than the rule (i.e., its structure and nature) itself. In the case of computer-aided sustainable design, the framework for sustainability is merely a pool from which specific elements and measures can be drawn to specify a specific relationship between a rating standard and a design model. This relationship constitutes a process of evaluation. As rating systems or design models evolve new relationships similarly evolve. Moreover, the processes can be extended to the urban scale in an unlikely fashion using software that is not geared to urban scale modeling. Likewise, in the case of surface tessellations, it is local panelization, a rule-based process, that ultimately specifies the overall design and fabrication. In the case of parametric shape grammars simplification through localized data structures, manipulations and rules in the form of 'code' is the first step to making grammars practically accessible to users. In short, part way to raising the *i*-factor is to transition design from a 'modeling' to a 'programming' exercise through local data structures and local manipulations. The next and perhaps more difficult step is to make the transition seamless.

Acknowledgement The work reported here would be impossible without the contributions of my graduate students Tajin Biswas, Tsung-hsien Wang, Peng-hui Wan, Kui Yue and Varvara Toulkeridou.

References

1. Autodesk (2003) Building information modeling for sustainable design. www.federalnewsradio.com/pdfs/BuildingInformationModelingforSubstainableDesign-white%20paper.pdf. Accessed 10 May 2010
2. http://www.usgbc.org/LEED. Accessed 10 May 2010
3. Krygiel E, Nies B (2008) Green BIM successful sustainable design with building information modeling. Wiley, Indiana
4. Biswas T, Krishnamurti R (2009) Framework for supporting sustainable design. Innovations for building and construction. Europia, Paris, pp 373–386
5. USGBC (2006) New construction and renovation guide V2.2, Washington
6. www.gbca.org.au/green-star/rating-tools/green-star-office-design-v3-green-star-office-as-built-v3/1710.htm. Accessed 12 May 2010
7. Echols S (2007) Artful Rainwater Design in the Urban Landscape. J Green Build 2(4):103–122
8. www.rhino3d.com. Accessed 12 May 2010
9. www.grasshopper3d.com. Accessed 12 May 2010
10. www.microsoft.com/NET. Accessed 12 May 2010
11. Kolarevic B (2005a) Architecture in the digital age: design and manufacturing. Taylor & Francis
12. Kolarevic B, Klinger K (2008) Manufacturing material effects: rethinking design and making in architecture. Routledge
13. Pottmann H, Asperl A, Hofer M, Kilian A, Bentley D (2007a) Architectural Geometry. Bentley Institute
14. Pottmann H, Liu Y, Wallner J, Bobenko A, Wang W (2007b) Geometry of multi-layer freeform structures for architecture. ACM Trans Graph 26:1–11
15. www.arcspace.com/architects/foster/elephant. Accessed 5 April 2010
16. www.fosterandpartners.com/Projects/1027/Default.aspx. Accessed 5 April 2010
17. Hauer E (2007) Erwin Hauer: continua-architectural screen and walls. Princeton Architectural
18. Hadid Z (2008) Schematic design report for next-gene architecture museum. National Chiao-Tung University, Hsinchu
19. Woodbury R (2010) Elements of parametric design. Taylor & Francis
20. Shimada K, Gossard DC (1995) Bubble mesh: automated triangular meshing of non-manifold geometry by sphere packing. ACM third symposium on solid modeling and applications, pp 409–419
21. Kolarevic B (2005b) Performative architecture: beyond instrumentality. Routledge.
22. Stiny G (1980) Introduction to shape and shape grammars. Environ Plan B 7: 343–351
23. Stiny G (2006) Shape: talking about seeing and doing. MIT
24. Stiny G, Gips J (1972) Shape grammars and the generative specification of painting and sculpture. The best computer papers of 1971. Auerbach, Philadelphia, pp 125–35
25. Gips J (1975) Shape grammars and their uses: artificial perception, shape generation and computer aesthetics. Birkhaüser, Basel
26. Stiny G (1975) Pictorial and formal aspects of shape and shape grammars and aesthetic systems. Birkhaüser, Basel
27. Stiny G (1977) Ice-ray: a note on Chinese lattice designs. Environ Plan B 4:89–98
28. Stiny G (1991) The algebras of design. Res Eng Design 2:171–181
29. Stiny G (1992) Weights. Environ Plan B 19:413–430
30. Taylor RG (1998) Models of computation and formal languages. Oxford University Press
31. Müller P, Wonka P, Haegler S, Ulmer A, van Gool L (2006) Procedural modeling of buildings. ACM Trans Graph 25(3):614–623
32. Müller P, Zeng G, Wonka P, van Gool L (2007) Image-based procedural modeling of facades. ACM Trans Graph 26(3):9 (article 85)

33. Watson B, Müller P, Veryovka O, Fuller A, Wonka P, Sexton C (2008) Procedural urban modeling in practice. IEEE Comput Graph Appl 28:18–26
34. Weber B, Müller P, Wonka P, GROSS M (2009) Interactive geometric simulation of 4D cities. Comput Graph Forum 28(2):481–492
35. Chau HH, Chen X, McKay A, de Pennington A (2004) Evaluation of a 3D shape grammar implementation. Design Computing and Cognition '04, Kluwer Academic, Dordrecht, pp 357–376
36. McCormack JP, Cagan J (2002) Supporting designers' hierarchies through parametric shape recognition. Environ Plan B 29:913–931
37. Yue K (2009) Computation-friendly shape grammars: with application to the determination of interior layout of buildings from image data. PhD Thesis. School of Architecture, Carnegie Mellon University
38. Aksamija A, Yue K, Kim H, Grobler F, Krishnamurti R (2010) Integration of knowledge-based and generative systems for building characterization and prediction. AIEDAM 24:3–16
39. Shape grammar implementation: from theory to useable software. http://www2.mech-eng.leeds.ac.uk/users/men6am/DCC10-SG-Implementation-Workshop.htm. Accessed 4 Nov 2010

Part II
Computer Science—State-of-the-Art

Dialectical Creativity: Sketch-Negate-Create

Tracy Hammond

Dialectical Creativity is the act of formulating a new concept through the original idea (the thesis), developing opposing contradictory ideas (the antithesis), and culminating on a more developed concretized idea that both negates and encompasses both the thesis and the antithesis (the synthesis). Sketching is a fundamental part of ideation. The act of performing ideation with an inherently abstract hand-drawn sketch, complete with messiness, allows the sketcher, through the misinterpretation of their own strokes, to evoke antithetical concepts, enabling the sketcher to quickly develop a creative synthetic idea. In the dialectical process there is a constant tension between creative change and the natural tendency to seek stability. Sketch recognition is the automated understanding of hand drawn diagrams by a computer, and can be used to both enhance creativity and/or idea stability. This paper discusses the Sketch Dialectic and its impact on the field of sketch recognition.

Introduction to the Dialectic

At its most simplistic essence, dialectics is the practice of using logical argumentation to investigate the truth of a theory or opinion. An early influential expression of this concept was in Plato's Socratic Dialogues, stemming from the idea of having two competing ideas going through a dialogue to eventually come to a state of better understanding. To gain an intuition for the choice of the term, notice that the term dialectic is related to the word 'dialect.' 'Dialectical', 'dialect', and 'dialogue' all stem from the same Greek word dialektos from dialegesthai, which means to "converse with each other," from dia- "across, between" + legein "speak".

However, this definition consisting of a purely logical argumentation does not get to the heart of the dialectic method, which requires the examination of both the original idea and the idea's negative to come to a synthetic conclusion. In Plato's Socratic Dialogues [2], Socrates was a contrarian and provided or encouraged the development

T. Hammond (✉)
Texas A&M University, Texas, USA
e-mail: hammond@cse.tamu.edu

J. S. Gero (ed.), *Studying Visual and Spatial Reasoning for Design Creativity*,
DOI 10.1007/978-94-017-9297-4_6

of an alternative hypothesis for any concept, trying to prove each idea wrong. The Socratic Method uses repeated oppositional questioning and the removal of hypotheses to produce an improved hypothesis. Socrates says that he never gives birth to any ideas (*Theaetetus* 148e–151d) [2], but rather uses repeated questioning to see if the ideas hold merit. Socrates himself would say that he has no ideas of his own, but he is just good at testing existing ideas. In this manner, Socrates is providing the antithesis for any idea.

These practices anticipated the Hegelian concept of the dialectic. The Hegelian dialectic further develops the dialectic process into a tri-phase model to explain the process of human history and the creation of truth through the tri-phase model of thesis-antithesis-synthesis. The dialectic works in a spiral-like repetitive process towards an ideal state, where each synthesis becomes the thesis for the cycle of dialectic idealism.

Although the Hegelian dialectic is associated with the terms thesisantithesis-synthesis, Hegel himself attributed these terms to Kant [10, 16], and instead used the terms abstract-negative-concrete and/or immediate-mediated-concrete, preferring these terms because he said that the antithesis implies that the negative comes from outside of a thing, whereas the negative comes from within and is inherent to the thing itself.

While the dialectic is associated with a quest for idealism, truth, and a concept of evolution and progress, a postmodern interpretation of the dialectic, and an abandonment of the concepts of evolution, progress, and truth, may lead us closer to the Deleuzean concept of lines of flight, involution, and becoming. With Deleuze, rather everything is inherently instable, and inevitably finds itself in a period of transformation.

The Deleuzean concept of becoming is differentiated from the dialectical process in many ways, two of which are significant here (a) its abandonment of the concepts of evolution towards and ideal and (b) its abandonment the negative and/or the antithesis. The dialectic is primarily concerned with progress from one point of history to another, with a formulation of betterment. In Plato's Socratic Dialogues, Socrates is trying to see how close he can get to an ideal, and attempts to approach that space through the examination of the negative. Contrastingly, Deleuze abandons this concept of the negative and progress towards an ideal. For Deleuze, all states and spaces are naturally and inherently unstable. For Deleuze, all objects go through the natural and repetitive process of: *I am here, yet "here" will inevitably self-destruct, and I will have to find a new place*. Deleuze states: "The term we would prefer for this form of evolution between heterogeneous terms is 'involution', on the condition that involution is in no way confused with regression. Becoming is involutionary; involution is creative. To regress is the move in the direction of something less differentiated. But to involve is to form a block that runs its own line 'between' the terms in play and beneath assignable relations" [4, 5]. The Deleuzean neoevolutionaryism concept of creation is both a movement of things away from other things without direction and without the Darwinian concept of progress, and a destruction of boundaries of what we thought we saw in order to see a particular entity.

When discussing creativity, both the concept of idea evolution (in terms of the standard concept of the dialectic) and idea involution (in terms of the Deleuzean concept of becoming) lead us to perhaps find a better fit for the current instability. The next section will discuss how the model of creativity fits both the traditional concept of progress as well as the postmodern concept of resettling in a place of inherent instability.

Dialectical Creativity and the Sketch Dialectic

Dialectical Creativity is the act of formulating a new concept through the original idea (the thesis), developing opposing contradictory ideas (the antithesis), and culminating on a more developed concretized idea that both negates and encompasses both the thesis and the antithesis (the synthesis). Traditionally, the Hegelian dialectic and Deleuzean becoming concern itself primarily with the evolution or involution of human society, contrastingly, dialectical creativity refers to ideation or the maturation of an idea.

Dialectical Creativity

Sketching is a fundamental part of ideation. The act of performing ideation with an inherently abstract hand-drawn sketch, complete with messiness, allows the sketcher, through the misinterpretation of their own strokes, to evoke antithetical concepts, enabling the sketcher to quickly develop a creative synthetic idea. The sketch is critical to creativity. In discussing or encouraging creativity, the pen and paper are quick to produce themselves. When experimenting with an idea, designers are quick to create a sketch to test out and manipulate an idea. Brainstorming is the dominant method for group creativity. More concrete design methods, such as CAD systems, are prevalent, but they are readily abandoned during the creative process. The Sketch Dialectic is the process of using Dialectical Creativity within the framework of sketching.

What is it about the sketch, the pen, and brainstorming that makes them so fundamental to creativity? In the dialectical process there is a constant opposition between creative change and the natural tendency to seek stability.

In the Hegelian dialectic version of Dialectical Creativity, the sketch is used to help designers' progress to an ideal solution that they are striving towards. The Encyclopedia of Creativity states: "In a dialectical process, the double functions of creativity are exercised. The positive function of creativity generates and constructs new concepts one after another. The negative function of creativity destroys preconceptions, displaces concepts, and breaks mental sets that would block imagination. Concurrent to the process of affirming new concepts, old concepts are being negated" [28]. Yan and Arlin continue to emphasize that any idea can be examined through the dialectical context: "When the process of dialectic is put in the context of a dialectical worldview, the creative nature of dialectical thinking would be magnified. From a dialectical perspective, every 'thing' is inherently contradictory with no limitation to the number of contradictions for each thing. In fact, each thing, as a part of a whole and also the whole of many parts, has multiple properties. As the thing interacts with a broader milieu, the number of properties multiplies and the nature of the properties changes. As one property has inherently on set of contradictions, one thing with multiple properties would have multiple sets of contradictions and, therefore, multiple corresponding resolutions. In this light, the development of a creative process is dialectical."

> In this dialectical process there is a constant opposition between creative change and the natural tendency to seek stability. This constant opposition creates a discomfort zone from which new and better ways of representing reality continually emerge. The overall process of dialectical thinking is, therefore in essence a process of self-perpetuating renewal and of self-perpetuating advancement. Some of the highest forms of creativity appear to be dialectical in nature. They often involve processes such as combining and recombining ideas, searching for complementarity, and coordinating multiple perspectives. Arlin in 1990 also noted that being dissatisfied the status quo, seeking new ways to formulate old problems, and noticing discrepancies unnoticed before are elements of intuition linked to the creative processes that are dialectical in nature. In fact, each of these processes of creative thinking can be regarded in part and parcel of problem finding, a concept originally used in the definition of a fifth/postformal stage in cognitive development. [28]

A sketch is a natural component of dialectic creativity. Sketching is inherently abstract, inherently imperfect, and we can harness the imperfections of a sketch to highlight negatives, which help to develop a concrete idea. The sketch allows you to remain in the abstract phase while negative are harnessed.

Deleuzean Becoming

The use of sketch within ideation also fits well to the Deleuzean concept of becoming. When we sketch, we don't draw what we see, we draw what they know. The sketch is inherently a messy incomplete drawing, and in interacting with the sketch, we complete things that aren't there. Inevitably our perception is faulty, and thus concept instability is going to develop.

Brainstorming is a free-flowing of ideas from one to the next. Sketching allows you to jot ideas informally on paper, but see them visually. A sketch provides you with the opportunity to explain or prove something to yourself. The pen and paper provide static immutable objects that are much more inflexible than a CAD system which allows you to more easily move around virtual objects. Why would these inflexible objects be preferred to mutable virtual objects?

I argue that it is not their immutability that makes a paper sketch more desirable. Rather, it is the fundamental messiness that is at its core. Sketches are, by their very nature, imperfect, and open for interpretation. When looking at a sketch, a designer has to interpret the sketch, has to make meaning out of the sketch, has to translate the sketch from abstract into reality. It is this internal human translation that breeds creativity. How does this internal translation breed creativity? A sketch breeds creativity through misinterpretation of its original intention. This idea of internal human translation or mistranslation of ones own ideas is supported by the idea that creativity in individual brainstorming (or free-writing or mind mapping) can be superior to that produced by group brainstorming [11].

Creativity and Sketching

The importance of sketching to enhance creativity has been well studied. Shah et al. has developed a concept called C-Sketch (collaborative sketch) that encourages creativity by having people trade sketches, and many of the creative ideas actually come from the previous sketch being misinterpreted. Researchers such as Shah, Jansson, and Smith have studied the inherent imperfections of a sketch, showing that the abstract sketch is often misinterpreted, and that those misinterpretations actually lead to greater creativity and better designs by helping to remove design fixation [18, 25].

It has been shown in architecture that having an abstract sketch is pivotal to removing design fixation on the part of the customer. Architects will often work in a design system, but when they show their designs to clients, client will often choose not to offer any suggestions to the drawings, rather they see them as concrete, something that they have to either accept or reject. However, when they offer clients instead, a hand-sketched diagram, clients take the opportunity to see and identify what is wrong in the design to come up with a pleasing solution. As such architects use programs to purposefully sketchify their design. Likewise, it is pivotal for engineers that they be able to sketch out their designs with a paper and pen in order to be able to properly formulate their ideas [23].

This paper describes several items of research that we are undergoing in the Sketch Recognition Lab and their relation to dialectical creativity in terms of the Hegelian dialectic, which may better praise usefulness and effectiveness concepts of creativity, and in terms of the Deleuzean becoming, which may better praise the concepts of novelty and innovativeness concepts of creativity.

The Sketch Dialectic and Sketch Recognition

The Encyclopedia of Creativity defines the dialectic as composed of three parts: "Thesis: In establishing a thesis, one is required to formulate a concept for analysis. Antithesis: In producing a contradictory view of the thesis, one is in fact generating an alternative concept that is in conflict with or in opposition to the original concept. Synthesis: To integrate the contradictory concepts into a dialectical whole, one would have to construct a whole new concept that is neither p nor q." [28] The dialectic is a constant struggle between enhancing creativity and the stabilization of that idea.

Sketch recognition is the automated understanding of hand drawn diagrams by a computer, and can be used to both enhance creativity and/or idea stability both in the Hegelian dialectical sense and in the Deleuzean Becoming sense. This section discusses the Sketch Dialectic and its impact on the field of sketch recognition.

Enhancing the Negative to Encourage Artistic Production

Perhaps the most obviously related project going on in our lab is that of our Dialectical Sketching Systems that purposefully maladapt and transform a user's sketches to the stroke style (using gestural stroke features) of an alternative persona to provoke idea instability and encourage creativity.

Techniques for sketch recognition usually fall into one of three camps: gestural, appearance-based, or perception/geometry-based. Gestural recognition focuses on the path drawn, including personal drawing-style information such as the curvature, timing, and pressure information. Although gestural recognition methods are usually used to define sketch-commands to the user, we can also use gesture recognition features to define a user's natural style of drawing in a freeform sketch environment. Each sketcher has a personal drawing style in the same way that each person has an identifiable personal handwriting. Gestural sketch recognition features can be measured to create a persona for a particular person's drawing style. This difference between this program and a standard filter is that a filter is designed to produce a desired effect, whereas the personas are designed to negate the desired effect.

We have developed a method of enhancing the negative to encourage artistic production by first measuring the gestural style of a particular person, then modifying the stroke to match the features of an alternative gestural style. The author assigned this task to her undergraduate Human Computer Interaction class. The students made seven different instances of this concept, producing interesting results. A local artist showed significant interest in the programs, and when asked to test one out for 5 min, he ended up glued to the program for over 2 h at his first sitting, leaving only when he was forced to by prior commitments. Despite several bugs being present in the first version, he emphasized that he wanted to use the program in his art class next fall. He stated that he felt that using the program would help learning artists to produce art more freely and confidently; he strives to have his students produce confident lines, owning the line and confirming each lines correctness even if the line was not the line that was originally intended. His favorite persona was that of the 'drunkard' which most significantly altered his original strokes, but he felt strongly that he could still 'see' his strokes in the seemingly random mess of strokes. Figure 1 is Andy Warhol inspired progression of a Campbell's Soup can.

Usually when you put a pen down on a paper or digital computer, the result is as expected, with the mark following your pen. What if the marks following behind

Fig. 1 An Andy Warhol inspired progression of a Campbell's Soup can

were not yours, but instead those of another personality? Your calm strokes translate to someone's who is anxious and panicked, or vice versa. What effect does that have on creativity? Mapping a user's sketched input to a predefined persona can lead to interesting forms of interaction. Sketch modification using personas turns a user's input into a dynamic sketch slightly beyond the user's control. This induced modification pushes the user out of her comfort zone and, by altering her work, can produce novel interactions with, and responses to, the computer. A user may find that she is entertained or possibly frustrated by the computer's alterations of her work, Figs. 2, 3, 4, 5 and 6.

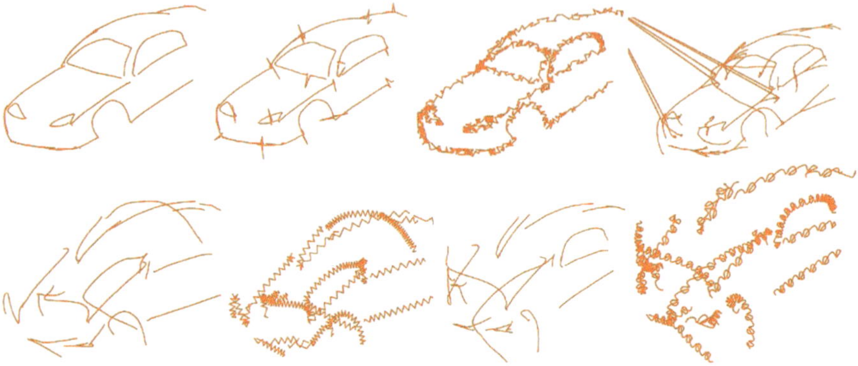

Fig. 2 Progression of a car

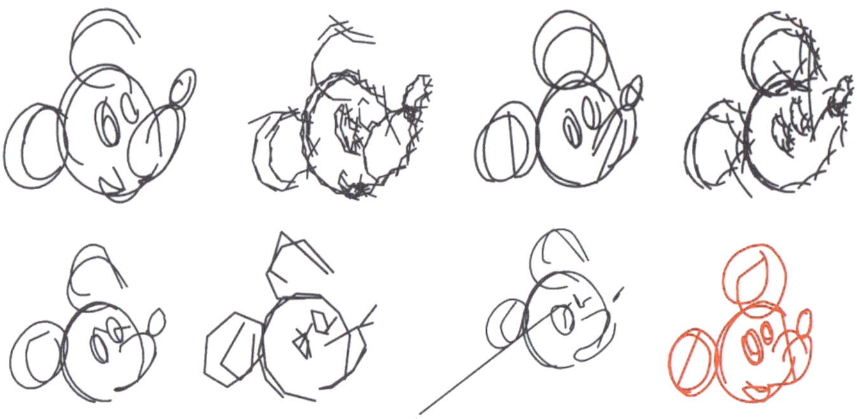

Fig. 3 Progression of a mouse

Fig. 4 When does the drawing cease to be Mickey? What about the Nike sign?

Fig. 5 Glasses. This is reminiscent of the "Choose Your Own Adventure" stories from childhood (a Rhizomatic narrative?). At each step along the way multiple lines of flight are possible

Fig. 6 Images produced from a version of the system that changes strokes in real time, forcing users to adapt to the altered versions of their strokes

Teaching Art and Creativity

The above system can be used to help teach art and creativity through the enhancement of the negative. In the case above, the negative comes from the computer system. A more formalized artistic instruction system could use sketch recognition

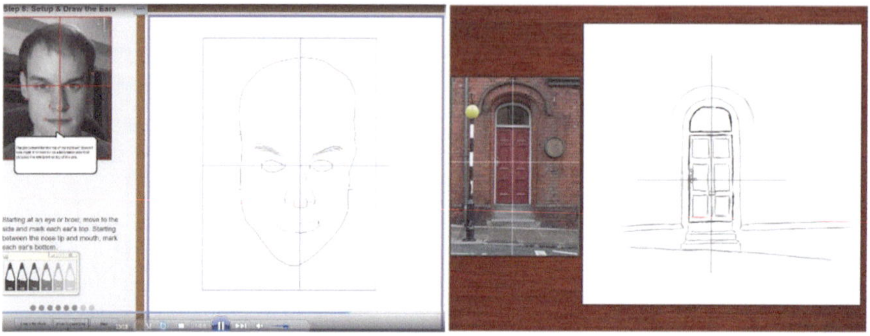

Fig. 7 The *iCanDraw* system teaching a user how to draw a face and an earlier version teaching perspective drawing of a door

techniques to enhance the creation of the synthetic. Interestingly enough, people don't draw what they see; they draw what they know. This is why traditional vision algorithms work so poorly on hand-sketched data. This is another reason why a sketch is so helpful to people because the contrast between the hand-sketch drawing and the actual object has to be rectified in their mind to create a synthetic ideal that is neither the original object nor their hand-drawn interpretation of the object. We can use dialectical creativity to explain the internal concept of a sketch: Thesis: The object being sketched. Antithesis: The sketched object. Synthesis: The internal perception of the object that both combines and negates both the original and the sketched object.

The goal in many artistic drawing programs is to get people to be able to sketch out the synthesis. As such the first step in such a program is to teach people to sketch the thesis (since what they sketch is already the antithesis) to that they can eventually sketch out the synthesis, Fig. 7.

The Sketch Recognition Lab has created the *iCanDraw* [7] program that uses sketch recognition techniques combined with vision and artificial intelligence techniques to teach users how to draw what they see (the thesis), as opposed to what they know (the antithesis). The program aims to teach them hand-eye coordination to provide them empowerment over the pen tool and the original object itself. The *iCanDraw* system teaches users how to draw a human face accurately. The system takes in a photograph that a user would like to learn how to draw and performs face recognition on that photograph to obtain a template of the face. The system then teaches the user to draw the face in a step-by-step manner using the process defined in the book 'Drawing on the Right Side of the Brain' [8, 9], and uses sketch recognition to provide real-time feedback and to provide an overall evaluation metric that compares the a face recognition produced template of the final sketch to the original photograph's template.

The above system teaches users how to draw what they see rather than what they know, but this is only the first step in teaching drawing, artistry, and creativity. By combining what the users see with what they know, sketchers can create a synthetic

Fig. 8 Common caricatures found on the internet given an example

concept that is negates both what they see and what they know and that has greater creativity than either idea alone. Figure 8 shows common caricatures found on the internet given an example.

Our goal is to be able to aid users to create a synthetic sketch that combines what they see and what they know in a way that is creative. As a first next step, we plan to aid users in the creation of vision-based and personality-based caricatures. In the vision-based caricatures, we will compare the face to the ideal (most common) human face and focus on what is most different from this ideal face. Are the eyes too far apart? Are the ears or nose too big? Then the caricature would enhance this negative to make this even more pronounced. We can use face recognition, sketch recognition, and mathematical models of the face to automatically determine possible variations for caricatures as well as automatically guide and teach the user how to identify variant features and draw their exaggerated caricature counterpoint. In the personality-based caricatures, we will merge a human face with an animal, either chosen by the user or automatically searching for shared adjectives used in describing both the animal and the person.

Collaborative Sketching

Collaborative sketching has proved to be beneficial to removing design fixation and causing improved ideation when using the Modified 635method of C-Sketch [25]. The effects of electronic sketching have been studied by [6]. Distributed sketching can prove to be a necessity in many circumstances as technological advances continue as more and more people choose to work in a distributed fashion, Fig. 9.

We created a collaborative sketching system called CoSke [3] that allows users to sketch collaboratively. Strokes appear in real-time to users as the user proceeds to lie ink down on the page (as opposed to at the competion of a stroke as is the

Fig. 9 The sketches produced by the four groups, shown in the order that they were created. The highlighted sketches were drawn in an isolated context

case in the recent Google Docs sketching program). We tested the program on 12 users divided in to four groups of three. Each group was asked to sketch a dragon, a bowl of fruit, and a house, in that order. Each sketch was performed in a different circumstance (with the circumstance order varied from group to group), either on a traditional piece of paper using colored pencils, using the CoSke system where all users were on individual computers but in the same room (so they could talk to one another), and using the CoSke system where they were placed in separate rooms so they were unable to speak to one another.

The participants were asked which method of interaction they preferred as well as the advantages and disadvantages for each. Disadvantages mentioned on the digital method were always because of technical issues, such as slow down, mis-functioning eraser, no undo, hard to draw straight lines, etc. Contrarily, the disadvantages of paper always included a comment implying that paper impeded the participants' ability to collaborate. Although none of the question specifically addressed the question of the physical interaction of paper sketching, 10 out of 12 users, including each user who preferred the paper interface, specifically commented on how paper affected their ability to collaborate because of the shared physical space.

Given that users felt obliged to mention this perception despite the lack of a specific question suggests the importance of further research on the following questions: Does paper impede sketch-based collaboration? Do user's need their own 'space' in order to collaborate effectively? Below we list the related comment from each user and their paper or digital preference:

U1: "I believe the paper interface didn't let multiple users to perform at the same time [in] the collaborative environment" (preferred digital) U2: "Pencil and paper was fun, but [it was] hard to draw at the same time." (preferred digital) U3: "I like to have control over my own drawing tool. The pencils were easier to control than the pen. However, it is more difficult for a group of people to draw with the same picture on a single sheet of paper collaboratively." (preferred digital) U4: "Drawing on [the] same piece of paper requires more cooperation." (preferred digital) U5: "The digital method was easier because you didn't have to worry about the physical constraints of the other's hands. You could draw in the same spot[more easily." (preferred digital) U6: "Both were easy to use, but we had to wait for each other on the paper one, because there wasn't enough space."(preferred paper) U7: "It was easier to use digital because I don't have to share the stylus as opposed to sharing color pencils." (preferred paper) U8:"[Paper provided an] opportunity for verbal and physical interaction." (preferred paper) U9: no related comment (preferred digital) U10: "Paper was easier, except that some people had to draw upside down, which was a huge benefit of the digital method." This user also mentioned that their group did not collaborate and all drew on their own parts of their paper. (preferred paper) U11: Digital was easier because we didn't have to awkwardly move our hands to make room for others. (preferred digital) U12:no related comment (preferred digital).

Combining speech with sketch proved to be pivotal for effective collaboration. 9/12 users preferred the collaboration with speech, and comment such as: "In different rooms there was no communication … lack of communication was difficult. I ended up drawing over people once or twice." were quite common. Interestingly, the three who preferred interacting without speech also noted that the lack of speech impeded collaboration: U: "Drawing without knowledge of what others were supposed to do, without set roles made it better." U: "There was no need to argue or communicate with others, we just drew. Everyone was working by themselves." U: "Isolated drawing made me feel like I was interacting with a language I knew nothing about and yet spoke more naturally than English… Interaction is very hard to coordinate when isolated, but once immersed, it becomes constant."

Looking at the figures produced through the different methods, no one method seems to produce overwhelmingly more creative solutions. Rather, it's possible to generalize that each method performed similarly. Creativity seemed to increase over time, probably because the participants became more comfortable with each other as a creative group.

To Beautify or Not to Beautify within Traditional uses of Sketch Recognition within Design

The Sketch Dialectic implies that in order to maintain the creativity-inspiring functions of a sketch, that when performing sketch recognition it is critical that we need to leave the sketch in its original strokes form. If we allow the sketch recognition to translate hand-drawn objects directly into a concrete form of a human's intention, just as in the case of working directly with a CAD system we can loose the natural creativity inspiring function of a sketch. This is not to say using sketch recognition in this automated translation method is not without merit, as there are many cases when interacting with a system using a pen is more simpler and more intuitive than interacting with a mouse and palette system. Immediate translation of pen strokes to their concrete intended meaning can significantly improve human computer interaction paradigms in many instances. However, while it improves the interaction paradigm, it does not increase creativity, since users are working with ideas that exist in well-defined and concrete terms, rather than interpretable ideas. I argue that it is in this act of internal interpretation that humans become more creative in that construct.

Sketch recognition does not have to deter creativity. It is only in the act of automatic concretization does sketch recognition hinder creativity. We can still use sketch recognition to understand user's diagrams while allowing them to maintain creative inspiration. In the design process, imagine if a user could draw and interact with their original strokes, while all the while the strokes are being translated simultaneously into the computer. The user can see and misinterpret their sketches in order to create these lines of flight. However, having their strokes recognized by the computer creates a number of obvious benefits.

The most repeated benefit is that upon finishing their sketch, designers do not have to repeat their final ideas into the computer for use within a CAD system. The computer can automatically interpret the sketch finalize the idea, and synthesize the concept. Usually this synthesis provides the added benefit of seeing this design in action [Gro96], such as a mechanical part in motion [1, 13, 21], generated code [14], mathematical solution of a problem [LZ04], or 3D walk through of a 3D space [20] Users get to work in the way that is most natural to them, significantly improving their interaction experience while still providing the benefits of a CAD tool and the creativity of a piece of paper. Users can get immediate feedback as to whether their design has merit. A blank piece of paper encourages creativity much more than a computer-automated drawing system does. Brainstorming is most often done with a pencil and paper because it does not constrain input. A constrained CAD drawing is easily recognizable by a computer. Graffiti provides a language for drawing shapes, but this language has to be learned. A human can understand and comprehend the drawings of another person, without needing to constrain the drawing style of the sketcher. Sketch recognition allows a computer to understand and automatically process natural hand drawings, Fig. 10.

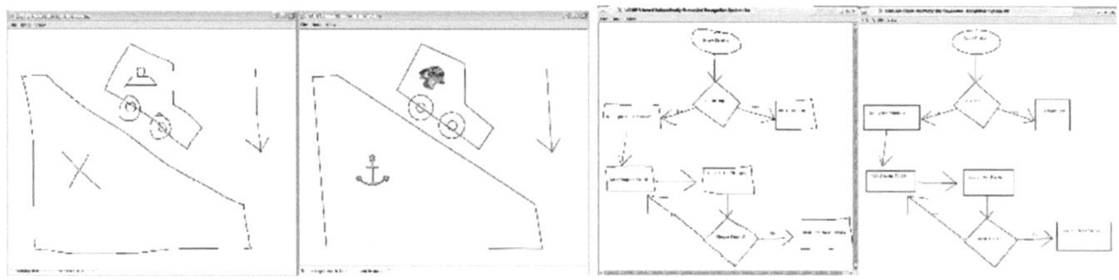

Fig. 10 Recognized diagrams of a mechanical engineering and a flowchart diagram

The benefits of sketch recognition in education are plentiful. Instructors and students sketch an abundance of graphical diagrams throughout the educational process. Graphical diagrams are particularly important and prevalent in Physics, and used throughout the learning process. However, correcting these diagrams proves difficult and time-consuming. Oftentimes, these graphical assignments or tests are omitted for these reasons. This is unfortunate as pedagogical studies suggest that not only does testing aid in learning, but that including testing in the educational process is more effective than test preparation alone [RK06]. Roediger explains that students remember more of what they learned when alternating only two study sessions with two testing sessions rather than by having four study sessions. Roediger also describes how early feedback after testing increases the percent of the material learned. Sketches can help in conceptual understanding, and instructors use sketches to communicate ideas [17, 27].

The drawing of these diagrams and automatic feedback is valuable to the student's learning process; however because the diagrams are often time-consuming to grade, the teacher may omit the drawing requirement and opt for a multiple choice testing scheme. Automatically correcting and understanding students' graphical diagrams will provide immediate feedback to the student and to the teacher about the student's understanding. The effect of this learning technology may encourage more instructors to include more graphical diagrams in their set of test questions, because previously time-consuming hand-correction of graphical diagrams can now be done automatically. Sketch recognition can help students to form a better conceptual model of the material, this research allows students to view the same problem in both modes, integrating creative and functional thought by allowing them to hand-draw images, and then providing them with immediate simulation feedback. Particular educational advancements include the ability to:

1. Provide active learning of concepts with a new interactive learning process,
2. Combine creative visualization (by providing students with drawing freedom by allowing them to hand-draw their diagrams as they would naturally) with functional visualization (simulating their own hand drawn diagrams along with enhanced editing possibilities),
3. Enhance learning by providing immediate feedback of homework, classwork, and test question answers,
4. Save instructor resources by automatically correcting student graphical work,

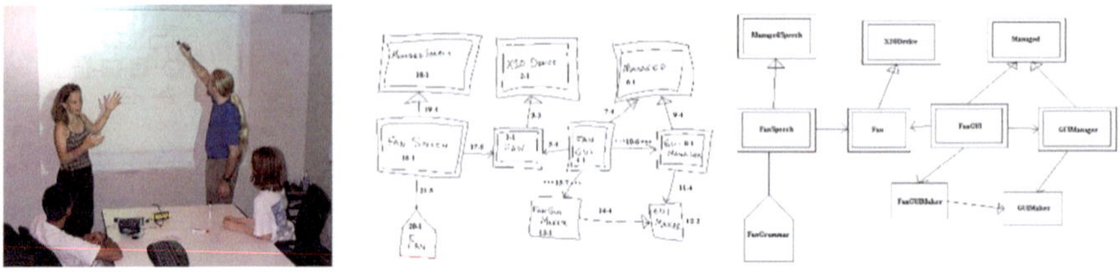

Fig. 11 Image of the system being used in a design meeting, a UML diagram produced, its cleaned up interpretation, and the ranking of the creation of each object and link

5. Provide instructor awareness of student comprehension by reporting and collating student results to the instructor,
6. Prevent cheating by generating different questions from semester to-semester or even student-to-student.

The benefits from the above scenarios come from the recognition of a finalized diagram. However, online recognition can provide many benefits to the creative process. An interpreted diagram can be manipulated and edited in a way that hand drawn sketches cannot. Perceptual grouping can be used to identify which items should be moved together [24], providing intuitive editing and diagram manipulation that has a significant advantage over static pen and paper diagrams that must be erased and redrawn. Additionally, because the diagram is recognized in real time, the designer can get feedback at any stage along the process to see if the concrete version of their current diagram would hold merit or work in the way that they were supposing. For instance, a civil engineer might want to first check that a particular beam can withstand the necessary force before building an entire bridge based on that assumption. Computers can be used along the process to provide feedback about parts of the design process.

Likewise, a computer that understands the sketch as it unfolds can watch the design process and automatically record design rationale based on the ever-changing sketch. In the work on UML class diagrams, we created a system that users could use in a software design meeting to automatically capture and document design rationale. The software designers designed an API using hand-sketched UML class diagrams in an intelligent room where both the white board and the videotaped session of the meeting was recorded. The hand-drawn UML class diagrams were understood in real time and the creation of editing of these class objects were used to automatically index the recorded meeting for later reference. Users could ask the system, "What were we discussing when we decided to create this class?" Significant events could be ranked and identified, so that users could also ask the system to "provide a history of the five most significant points in the design process." [15], Fig. 11.

Conclusion

This paper discusses and defines Dialectical Creativity and the Sketch Dialectic and its impact on the field of sketch recognition. In the dialectical process there is a constant opposition between creative change and the natural tendency to seek stability. The imperfections of hand-drawn sketches inherently aid the production of antithetical concepts that can help increase creativity. Sketch recognition is the automated understanding of hand drawn diagrams by a computer, and can be used to both enhance creativity and/or idea stability.

Acknowledgements This work was supported in part by the DARPA Computer Science Study Panel. Much of the work described above was performed in collaboration with past and current graduate and undergraduate student members of the Sketch Recognition Lab at Texas A&M University, with a special shout out going to Daniel Dixon, Manoj Prasad, Jessica David, and Nic Lupfer. Additionally, thanks to Jennifer Mease, Christopher Swift, Stace Treat, Christopher Zebo, Tobias Tolland, and Kerre Cole for their help formulating the concept of Dialectic Creativity and the Sketch Dialectic.

References

1. Alvarado C, Davis R (2001). Resolving ambiguities to create a natural computer-based sketching environment. In Proc. of IJCAI '01, pp 1365–1371
2. Cooper BMA (1997). Plato: complete works. Hackett, Indianapolis
3. David J, Eoff B, Hammond T (2010) CoSke: An Exploration in Collaborative Sketching. Posters at the 2010 ACM Conference on Computer Supported Cooperative Work, Savannah, Georgia, February 6–10
4. Deleuze G (1986) Foucault. University of Minnesota Press, Minneapolis
5. Deleuze G, Guattari F (1987) A thousand plateaus: capitalism and schizophrenia. University of Minnesota Press, Minneapolis
6. Dennis AR, Williams ML (2003) Electronic brainstorming: theory, research, and future directions. In: Paulus P (ed) Group creativity. Oxford University Press, Oxford
7. Dixon D, Prasad M, Hammond T (2010) iCanDraw? using sketch recognition and corrective feedback to assist a user in drawing human faces. In Proceedings of Computer Human Interaction (CHI 2010), ACM Press, Atlanta, Georgia. April 10-17:2010
8. Edwards B (1998) Drawing on the right side of the brain. In Proc. of the SIGCHI Conference on Human Factors in Computing Systems, pp 139–140
9. Edwards B (1999) The new drawing on the right side of the brain. Penguin, New York
10. Fox M (2005) The accessible Hegel. Prometheus, Amherst
11. Furnham A, Yazdanpanahi T (1995) Personality differences and group versus individual brainstorming. Personality Individ Differ 19:73–80
12. Gross M (1996) The electronic cocktail napkin: a computational environment for working with design diagrams. Des Stud 17(1):53–69
13. Hammond T, Davis R (2002) Tahuti: a geometrical sketch recognition system for uml class diagrams. AAAI Symposium on Sketch Understanding (AAAI 2002), San Francisco, CA. July 28–August 1
14. Hammond T, Davis R (2004) Automatically transforming symbolic shape descriptions for use in sketch recognition, The Nineteenth National Conference on Artificial Intelligence (AAAI-04)

15. Hammond T, Gajos K, Davis R, Shrobe H (2002) An agent-based system for capturing and indexing software design meetings. Workshop on Agents in Design, August 28
16. Hegel GWF (1977) The phenomenology of the spirit, [Trans. A.V.Miller]. Claredon, Gloucestershire
17. Horn R (1998) Visual language: global communication for the 21st century. MacroVU, Bainbridge Island
18. Jansson DG, Smith SM (1991) Design fixation. Des Stud 12(1):3–11
19. LaViola Jr. JJ, Zeleznik RC (2004) MathPad2: a system for the creation and exploration of mathematical sketches. ACM Trans Graphics 23(3):432–440
20. Oh Y, Do EY-L (2004) Intelligent critiquing of design sketches. In: AAAl Fall Symposium, Making Pen-Based Interaction Intelligent and Natural. AAAI Press
21. Oltmans M, Davis R (2001) Naturally conveyed explanations of device behavior. In Proceedings of PUI '01, pp 1–8
22. Roediger HL, Karpicke JD (2006) The power of testing memory: basic research and implications for educational practice. Association Psychol Sci 1(3):181–210
23. Rose A (2005) Graphical communication using hand-drawn sketches in civil engineering. J Profess Issues Eng Educ Pract. doi:10.1061/(ASCE)1052-3928(2005)131:4(238)
24. Saund E, Mahoney J, Fleet D, Larner D, Lank E (2002) Perceptual organization as a foundation for intelligent sketch editing. In Sketch Understanding, Papers from the 2002 AAAI Spring Symposium, pp 118–125
25. Shah J, Vargas-Hernandez N, Smith S (2003) Metrics for measuring ideation effectiveness. Elsevier, Amsterdam, pp 111–134
26. Shum S, MacLean A, Bellotti V, Hammond N (1996) Graphical argumentation and design cognition. Hum Computer Interact 12(3):267–300
27. Tversky T (2002) What do sketches say about thinking? Sketch understanding, Papers from the 2002 AAAI Spring Symposium, pp 148–151,March 25–27 2002
28. Yan B, Arlin P (1999) Dialectical thinking: implications for creativethinking. In: Runco M, Pritzker S (eds) Encyclopedia of creativity, vol 1. Elsevier, Amsterdam, pp 551–555

Spatial Computing for Design—an Artificial Intelligence Perspective

Mehul Bhatt and Christian Freksa

Artificial Intelligence and Design

The significance and the paradigmatic relevance of Artificial Intelligence in Modern Design is intertwined with Herbert Simon's original articulation of the *Science of Design* [41, 42] and in the words of Baldwin [4], Simon's interpretation of design as a "*decision-making process under constraints of physics, logic and cognition*". This view of the scientific design process underlies much of what artificial intelligence has to offer by way of its formal representational and computational apparatus to the domain of design.[1] From a topical viewpoint, the knowledge representation and reasoning area within artificial intelligence has been the cornerstone of most formal AI inroads in so far as *problem-solving for design* is concerned. In the last two decades, several interdisciplinary initiatives comprising of computer scientists, engineers, psychologists and, designers have addressed the application of artificial intelligence techniques for solving problems that accrue at several stages of the design process: design conceptualization, functionality specification, geometric modelling, structural consistency & code-checking, optimization, collaborative (design) workflow management, design creativity, and a plethora of other issues.[2]

Situated within this AI-centric view of the science of design, we present our perspective on *spatial computing for design*. Strongly influenced by the need to

[1] Henceforth, by design we refer to spatial design in general, and in specific to architectural design, which we regard to be an instance of spatial design. By conventional design systems, we refer to computer-aided architectural design (CAAD) tools.

[2] The journal "Artificial Intelligence for Engineering Design, Analysis and Manufacturing" completed two decades of publishing in 2007 and its anniversary publication is a good overview of the area [15, 25]. A sketch of '*40 years of design research*' is available in [5]. The collected works of [1, 14, 17, 23, 26, 29, 34] are a rich source of reference and contextualisation.

M. Bhatt (✉) · C. Freksa
University of Bremen, Bremen, Germany
e-mail: bhatt@sfbtr8.uni-bremen.de

C. Freksa
e-mail: freksa@sfbtr8.uni-bremen.de

J. S. Gero (ed.), *Studying Visual and Spatial Reasoning for Design Creativity*,
DOI 10.1007/978-94-017-9297-4_7

formally define, model, and reason about (structural) form & (artefactual) function, our interpretation of spatial computing encompasses three aspects we regard as crucial:

- semantic modelling, spatial abstraction, & multi-perspective representation
- design analysis by inference patterns supporting diagnostic & hypothetical reasoning
- assistive feedback/communication with designers

1. The aspects deemed essential correspond to problems that accrue within a conventional 'iterative refinement by automated design assistance' work-flow, and are identifiable with respect to the *modelling–evaluation–re-design* phases in intelligent design assistance, for instance, as interpreted within a Function-Behaviour-Structure (FBS) [23, 26] model of the design process. With respect to the refinement work-flow, the basic research questions within the context of spatial computing include: Semantics: formal modelling of design requirements, and the role of knowledge engineering in that regard
2. Spatial abstraction: abstraction of CAD-based geometric information into the qualitative domain via the use of formal spatial representation and reasoning techniques
3. Qualitative spatial reasoning: the application of spatial consistency as a basis for checking for design requirement consistency
4. Hypothetical reasoning: the role of hypothetical reasoning (e.g., by abduction) as a means to support a diagnostic and recommendation function within a logic context
5. Assistive feedback: visualisation modalities as a means to interact & communicate assistive feedback with the designer

The above problem aspects have fuzzy boundaries and many interrelationships. However, the paper attempts to characterize each one of them rather independently via running examples. The paper is organized as follows:

The next section is an exposition of the philosophy that underlies our approach to spatial computing. The points raised condition the basic premises of our overall approach, especially our propositions on hypothetical reasoning for design. The next section provides an overview of the iterative refinement cycle in design. Here, we exemplify the key aspects of spatial computing for design vis-a`-vis the iterative refinement cycle. In the next section, we define spatial computing and present the issues that we deem to be within its scope. Key representational and computational modalities are discussed, and we also attempt to ground the earlier discussion with examples that further illustrate the agenda of spatial computing, and the problems that may be solved therein. This section can be considered to be our statement of the work-in-progress in spatial computing for design. Finally, we summarize and at the same time, also reflect upon some of the issues raised by Gero [24] in his statement of "*Ten Problems for AI in Design*".

The Philosophy of Spatial Computing, for Spatial Design

In architectural design, we are faced with structures in physical space. Much of the design considerations in architectural design are directly constrained by intrinsic properties of physical space. Unlike some abstract spaces, the dimensions of physical space are strongly interrelated. This has to do with the fact that the three spatial dimensions are of the same quality: an object that is long in the x-dimension will be long in the y- or z-dimension simply by changing its orientation; one does not have to change the (nature of the) object itself. In color space, for example, we cannot maintain such constancy by moving an object, as each dimension of color space refers to a different quality or feature. On the other hand, the number of spatial dimensions is limited to three; thus, in whatever ways we move objects in space, we will stay within the interrelation of three spatial dimensions.

Besides these intrinsic spatial constraints, we have physical constraints due to mechanical properties of physical objects. In particular, there is a correlation between length, width, and height due to mechanical requirements: longer objects need to be made thicker for maintaining stability properties; thicker objects may become larger to maintain proportions, etc.

Human perception also treats the three spatial dimensions in similar ways: for manipulable objects the perception of length, height, and width can be transformed by changing the orientation of the objects; for large-scale objects like buildings and mountains, the vertical dimension may not be perceived identically to the two horizontal dimensions.

The main message of these considerations is that physical dimensions are strongly inter-related and that physical space is severely constrained. This can be viewed as a strong limitation in comparison to abstract spaces in which arbitrary configurations of feature values and arbitrary transitions between them are conceivable. From the perspective of design, however, these constraints can be considered a great advantage, as they considerably reduce the space of design decisions.[3]

These considerations not only have implications on the spatial structures to be designed but also on the structures of design computers. Today's general purpose computers represent spatial entities and environments in the conceptual framework of unconstrained abstract spaces; thus, the intrinsic properties of physical space must be explicitly coded into the system to make sure physically realizable designs result from the computational process.

In other words, computation needs to be invested to reduce the set of conceivable designs to the set of realizable designs. This is not the case when the designer works directly with spatial models, as these maintain the spatial constraints inescapably.

[3] Hypothetical reasoning about designs focussing on *what could be* rather than *what is* benefits from rich on- tological characterizations along these lines. This is further elaborated on in Sect. 4.4). Also, see the treatment of aspectualization for architectural design in [7, 8].

We use the notion of spatial computing in a way that exploits the intrinsic constraints of spatial structures in such that only those structures will be generated that are realizable in physical space and that do not require a computational reduction from conceivable structures to physically realizable structures.

Assisted Iterative Refinement in Spatial Design

Spatial design as a problem-solving activity typically consists of the Conception—Modelling—Evaluation—Re-modelling cycle. Essentially, a designer in this case is considered to be an *agent of change*, who in the absence of any computational assistance, may be intuitively regarded as traversing a complex configuration space of possibilities, and selecting one course of action (guided by domain knowledge, expertise, cognitive capabilities, specialized requirements, aesthetic preferences and so forth) that produces a desired product/design.

A Design Task. As a basic use-case, consider an architect/engineer specialising in the design and development of building automation systems and smart environments. A typical design challenge would be:

> Design the layout of an office environment to satisfy structural and functional requirements that collectively aid and complement (and never hinder) the building's automation systems (monitoring devices, sensors, etc.), and which, by implication, facilitate the intended smartness of such automation systems.

From the viewpoint of the overall design requirements, aspects of this problem explicitly pertain to the functional aspects (e.g., security, privacy, building-automation, accessibility) of the space being modelled, structural code-checking with respect to building regulations, and also possibly specialized client demands. Some example requirements follow in (R1–R3):

R1 certain areas within a building/floor/room should (not) be trackable by sensing devices such as cameras, motion-sensors

R2 regional statutory requirements that stipulate structural constraints and other categorical specifications, e.g., disability access codes, design guides

R3 client specification: as much as possible, the operation of doors should be non-interfering with the functionality of nearby utilities/artefacts

Figure 1 is a schematization of the consistent and inconsistent models of the example requirements/scenarios in (R1–R3). The following aspects, marked as [1–4] in Figs. 1a and b, make the plans of Fig. 1 (in)consistent with respect to (R1–R3):

- the sensor/camera is placed at a place where a private area such as the washroom is within its range (No. 1)
- the operating space of the door of the wash-room interferes with the functional area of the wash-sink, and this arrangement is also not conducive, given disability access requirements (No. 2)

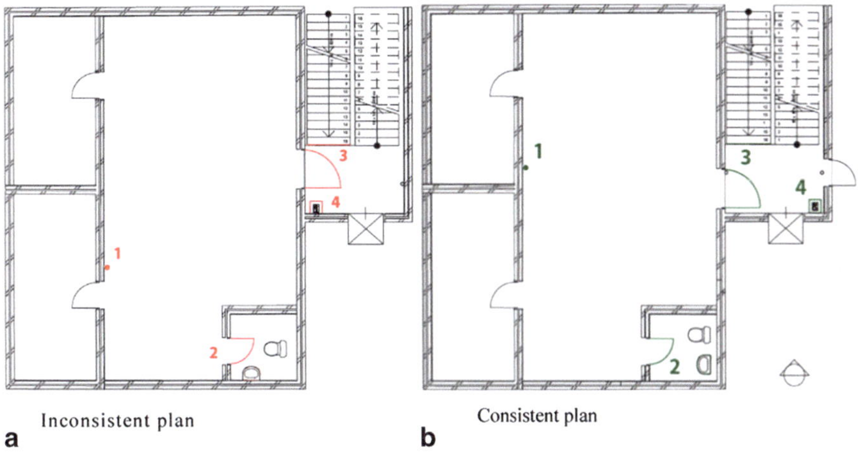

Fig. 1 Design requirements: example spatial interpretations. **a** Inconsistent plan. **b** Consistent plan

- the operation of the main entrance door interferes with the function of the telephone next to it, and from a structural viewpoint, is also not an ideal placement given its proximity to the staircase (No. 3, 4)

From the viewpoint of spatial computing, one may imagine the search space to consist of spatial configurations—topological, orientational arrangements—and the spatial trans- formations that are possible, e.g., with respect to a movement taxonomy, as the available actions that produce a re-arrangement. The objective of iterative refinement in general, be it automated or human, is to create consistent models that fulfill the requirements as they are conceived at design time. Albeit a bit limiting, for this particular case, the automation necessary to realize the re-configuration may be identified as a limited form of assistive spatial design intelligence that guides the designer toward a solution that meets the pre-specified requirements, such as those stipulated in (R1–R3).

Automated Design Refinement Figure 2 illustrates our interpretation of this process of iterative refinement, as it is applicable to the 'spatial computing for design' framework (Sect. 4) laid out in this paper. The following aspects of iterative refinement (A1–A3) are deemed crucial:

A1 *Modelling—Design Abstraction*: this aspect encompasses issues ranging from se-mantic specifications, taxonomic representations, qualitative abstractions of geometric models, and modularity of information representation

A2 *Convergence—Reasoning:* this aspect constitutes the various modes of inference that constitute the computational manifestations of the assistive design support

A3 *Assistive Feedback—Visualization:* this aspect constitutes mechanisms and modalities to provide diagnostic feedback and other forms of support within a conventional CAAD workflow

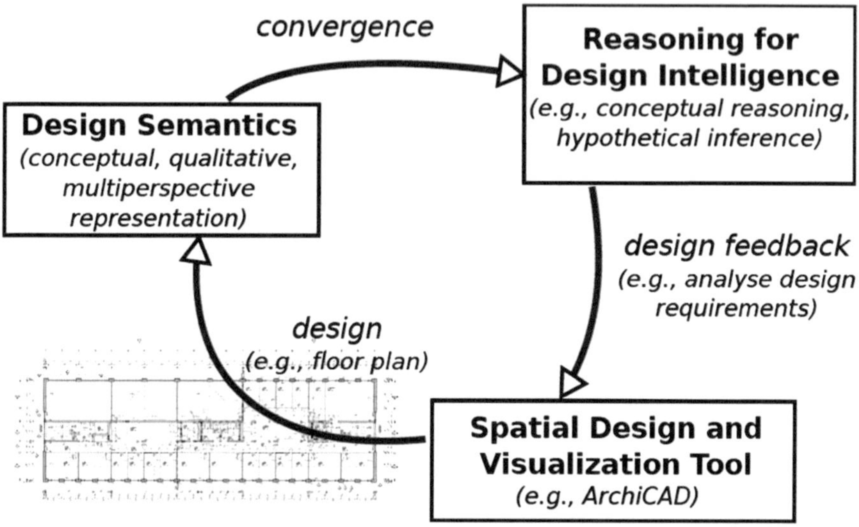

Fig. 2 Iterative refinement by intelligent design assistance

Indeed, the possibilities to broaden the interpretation of this manner of intelligent assistance are rather extensive, ranging within a wide array of techniques from the computing, cognitive, psychological, and aesthetic disciplines. Our preliminary focus in spatial computing is centred on spatial cognition, and is guided by the aim to formally and computationally understand the relationship between the "structural form" and "artefactual function" within the domain of spatial design. Further elaborations are presented in Sect. 4.

A Characterization of Spatial Computing for Design We characterize *spatial computing for design* in two ways: firstly, by the scientific questions that it must address from a representational and computational viewpoint and their relationships to the domain of artificial intelligence & design in general, and secondly, by the outcomes that a paradigm such as this is expected to produce. Spatial computing for design is defined as:

- that body of work that is concerned with the use of formal methods in knowledge representation and reasoning in general, and terminological and spatial representation and reasoning in specific, for solving problems in modelling (e.g., spatial semantics, modularity, requirement constraints) and validation (e.g., diagnosis, hypothetical reasoning) in the domain of spatial design
- that body of work whose aim is to develop the generic apparatus—application framework, methodology, tool-sets—that may be used as a basis of providing assistive design support within a conventional CAAD-based spatial design workflow

We now elaborate on the representational and computational aspects of the above definition.

Modelling Form and Function "*Form follows Function*" [44] and "*Ornament is Crime*" [35]—these two doctrines have been the cornerstones of the modernist tradition in engineering design.[4] Restricting the application of this doctrine to the domain of architectural design, the interpretation that it leads to is that the *structural form*, i.e., *shape, layout, connectivity*, of a building should be primarily (or more rigidly: *solely*) determined by its practical *function* or *purpose*. Much of the literature in the philosophy of design and architecture [48], and the ensuing debates thereof, have focused on the semantics of *functions* with respect to design artefacts and the causal link between *form* and *function*, stressing the question of whether or not form should, or indeed does, wholly or in part follows function.[5]

Structural Form and Artefactual Function in Spatial Computing: Spatial computing is primarily concerned with the issues surrounding the formal interpretation of the terms "spatial/structural form" and "artefactual function", in particular with respect to the interpretation of these concepts in the context of a CAAD-based workflow. This is crucial, since it is necessary to explicitly put these notions into practice by investigating what precisely does it mean to model *form* and *function* within an intelligent architectural design assistance system.

Example 1. Bremen (Germany) Building Code [12]:
(a). *Staircase/Treppen* (§ 35 (10), p. 24):

> Steps of a staircase may not be connected directly to a door that opens in the direction of the steps. There has to be a landing between the staircase steps and the door. The length of this landing has to have at least the size of the door width.

We note some examples:
Example 2. US Courts Design Guide 2007 [47]:
(b). Barrier-Free Accessibility (p 4–3):

> Courtroom areas used by the public must be accessible to people with disabilities. Private work areas, including the judge's bench and the courtroom deputy, law clerk, bailiff, and court reporter stations, must be adaptable to accessibility. While all judges benches and courtroom personnel stations do not need to be immediately accessible, disabled judges and court personnel must be accommodated

(c). Psychology, Culture and Aesthetics (p. 3–1, 4–4):

> The architecture of federal courthouses must promote respect for the tradition and purpose of the American judicial process. To this end, a courthouse facility must ex- press solemnity, integrity, rigor, and fairness.
> All architectural elements must be proportional and arranged hierarchically to signify orderliness. The materials employed must be consistently applied, be natural and regional in origin, be durable, and invoke a sense of permanence.

[4] Whereas Louis Sullivan articulated the relationship between of 'Form and Function', the original attribution goes to the eighteenth century Italian architectural theorist Carlo Lodoli.

[5] Dorst and Vermaas [20] present a critical review of the Function-Behaviour-Structure model. The discussion sheds useful insights about the nature of form-function.

> The height and location of the judges bench expresses the role of the judge and facilitates control of the court. Generally, the judges bench should be elevated three or four steps (21–24 inches or 525–600 mm) above the courtroom wall.

(d). Visibility (p. 3–2, 16–9):

> The entrance or entrance vestibule should be clearly visible and recognizable as such from the exterior of the building. The vestibule should be a minimum of 7 feet in depth and able to handle the flow of traffic at peak times.
> A duress alarm must be easily accessible and visible to all occupants.

Example 3. A Pattern Language [2]
(e). *Sunny Counter* (p. 16–918):

> Place the main part of the kitchen counter on the south and southeast side of the kitchen, with big windows around it, so that sun can flood in and fill the kitchen with yellow light both morning and afternoon

At this stage, we leave the readers with their imagination as to the formal interpretation of the above examples—some have a clear and well-defined spatial structure within a design, whereas others are only indirectly specifiable. Spatial computing in design should be concerned with the extent to which functional aspects such as those exemplified herein could be formally interpreted in strictly semantic and spatial terms; from a computational viewpoint, it is clear that adequate conceptual, spatio-linguistic and qualitative modelling techniques are necessary for representing and reasoning about *design artefacts* and *patterns* entailed by designer expertise.

Design Artefacts: Conceptualization and Formal Representation Spatial computing involves an interplay between the designer's conceptualization, the handicaps of the computational constructs of the design tool, and the limitations of the bridges that connect that conceptual with the computational: professional design tools simply lack the ability to exploit the expertise that a designer is equipped with, but un- able to communicate to the design tool explicitly in a manner consistent with its inherent human-centred conceptualization, i.e., semantically and qualitatively. Modelling for spatial computing in design has to be focussed on representation of *design semantics, artefactual modelling* capability and support for *multi-perspective modularity*.

Design Semantics An expert's design conceptualization is semantic and qualitative in nature—it involves abstract categories such as Rooms, Doors, Sensors and the spatial (topological, directional, etc.) relationships among them, e.g., 'Room A and Room B have a Door in Between, which is monitored by Camera C'. Whereas this example is rather specific, typical real-world constraints are mostly underspecified or fuzzy (e.g., see Sect. 4.1). Therefore, any vision of specialised spatial computing for design has to handle to the modelling of designer/design semantics in an explicit manner, e.g., using formal knowledge engineering constructs such as ontology modelling languages.

Spatial Artefacts A crucial aspect that is missing in contemporary design tools is the support to explicitly characterize the artefactual aspects, and the functional requirements ensuing therefrom, within a design. Semantic descriptions of

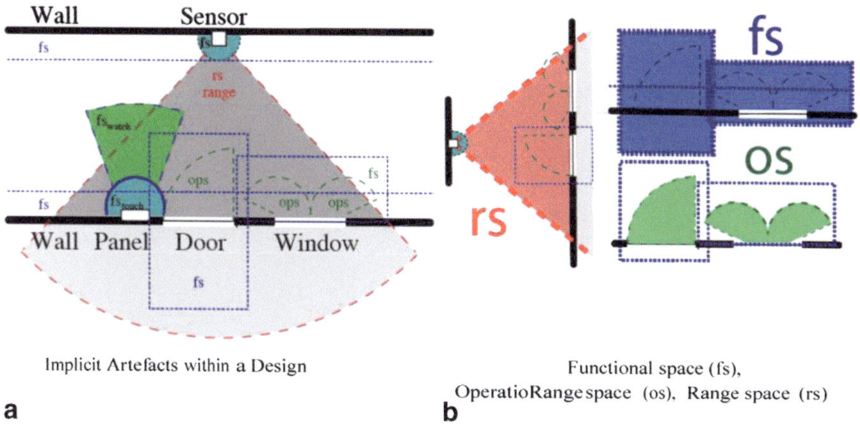

Fig. 3 Spatial artefacts are entities, which unlike regular spatial objects, do not have a physical manifestation in reality (or within a design), but need to be treated as such for all practical/reasoning purposes. (Illustration adapted from: Bhatt et al. [13])

designs and their requirements acquires real significance when the spatial and functional constraints are among strictly spatial entities as well as abstract spatial artefacts. For instance, although it is possible to model the spatial layout of an environment at a fine-grained level, it is not possible to model spatial artefacts such as the range space of a sensory device (e.g., camera, motion sensor, view-point of an agent), which is not strictly a spatial entity in the form of having a material existence, but needs to be treated as such nevertheless. In general, architectural working designs only contain physical entities. Therefore, it becomes impossible for a designer to model constraints involving spatial artefacts at the design level. For instance, consider the following constraint: 'the motion-sensor should be placed such that the door connecting room A and room B is always within the sensor's range space'. Bhatt et al. [13] identify three types of spatial artefacts:

A1 the *operational space* denotes the region of space that an object requires to perform its intrinsic function that characterizes its utility or purpose

A2 the *functional space* of an object denotes the region of space within which an agent must be located to manipulate or physically interact with a given object

A3 the *range space* denotes the region of space that lies within the scope of a sensory device such as a motion or temperature sensor, or any other entity capable of visual perception. Range space may be further classified into other categories, such as *observational space* (e.g., to model the concept of the *isovist*[6]).

Figure 3 provides a detailed view on the different kinds of spaces we introduced. From a geometrical viewpoint, all artefacts refer to a conceptualised and derived

[6] An isovist is the set of all points visible from a given vantage point in space and with respect to an environment [6].

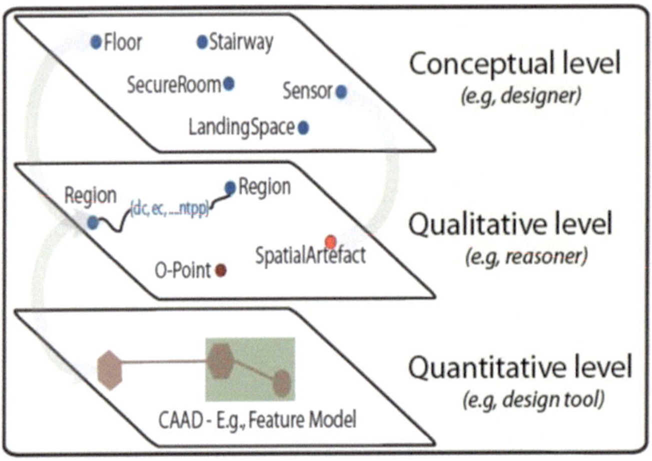

Fig. 4 Multi-perspective representation & modularity

physical spatial extension in Rn. However, they do differ from an ontological perspective and the manner in which their geometric interpretations in Rn are derived. The derivation of an interpretation may depend on object's inherent spatial characteristics (e.g., size and shape), as well as additional parameters referring to mobility, transparency, etc.

Multi-Perspective Semantics & Representational Modularity An abstraction such as a Room or Sensor may be identified semantically by its placement within an ontological hierarchy and its relationships with other conceptual categories. This is what a designer must deal with during the initial design conceptualization phase. However, when these notions are transferred to a CAAD tool, the same concepts acquire a new perspective, i.e., now the designer must deal with points, line-segments, polygons and other geometric primitives available within the feature hierarchy of the design tool, which, albeit necessary, are in conflict with the mental image and qualitative conceptualization of the designer. Given the lack of semantics, at least within contemporary design tools, there is no way for a knowledge-based system to make inferences about the conceptual design and its geometric interpretation within a CAAD model in a unified manner.

As an example, consider a binary relation 'connects' that links entities from the conceptual, qualitative, and quantitative levels of Fig. 4; a Floor at the conceptual level is abstracted as a Region at the qualitative level of a reasoner and as a ClosedPolygon thereby preserving the geometry at the quantitative level of a CAAD-based feature model:[7]

[7] The examples are illustrated using a scheme that is close to the so-called Manchester Syntax Horridge and Patel-Schneider [2008] for the description of ontological knowledge in the Web Ontology Language (OWL). The syntax 'M:C' represents a concept 'C' within particular ontological module 'M'. Formal descriptions for these examples may be found in [30].

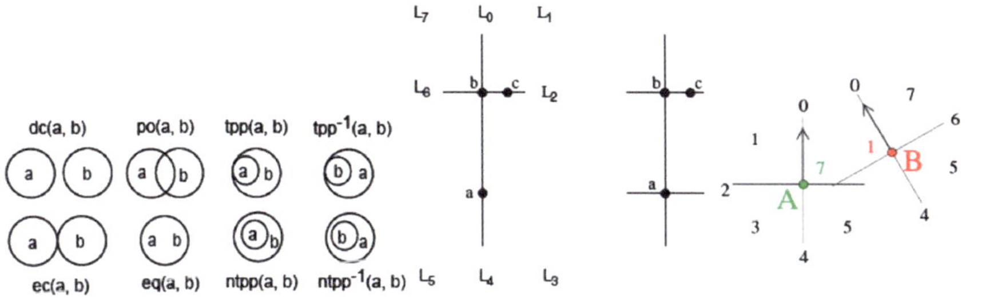

RCC-8 Primitives
a

SCC & DCC Partitioning
b

The $OPRA_2$ relation $A\ {}_2\angle^1\ B$
c

Fig. 5 Topological and orientation calculi

1	*BinaryLink: Domain:*	integration module: connects quantitative level: architectural feature
	Range:	qualitative level: functional structure
	Inverseof:	integration module: connectedby
2	*Class: subclassof:*	quantitative level: convexpolygon connects exactly 1 qualitative level: region
3	*Class: subclassof:*	conceptual level: floor connects exactly 1 qualitative level:region

Spatial Representation and Reasoning The field of Qualitative Spatial Reasoning (QSR) investigates abstraction mechanisms and the technical computational apparatus for representing and reasoning about space within a formal, non-metrical framework [18, 21]. Relational formalizations of space and tools for efficiently reasoning with them are now well-established [40]. In QSR, spatial information representation corresponds to the use of formal spatial calculi such as the Region Connection Calculus [39] (RCC), Single-Cross and Double-Cross Calculi (SCC, DCC) [22], Oriented Point Relation Algebra (OPRA) [37] (see Fig. 5).

Within spatial computing for design, the use of formal qualitative spatial calculi and conceptual design requirements serve as a link between the *structural form* of a design and the differing *functional capabilities* that it affords or leads to. Therefore, a very important goal in spatial computing is to formally and computationally investigate the link between structural forms, as denoted by specific spatial configurations of domain entities, and the behaviours/functions that they are inherently capable of producing with respect to a pre- specified set of requirements conceptually expressed by an architect or a designer.

Artefactual Constraints, Structural Form and Design Function Spatial artefacts such as those introduced in (A1–A3) are usable towards formulating functional requirement constraints for a work-in-progress spatial design. Constraints, emanating from the requirements such as in (R1–R3; Sect. 3) may need to be satisfied by a design:

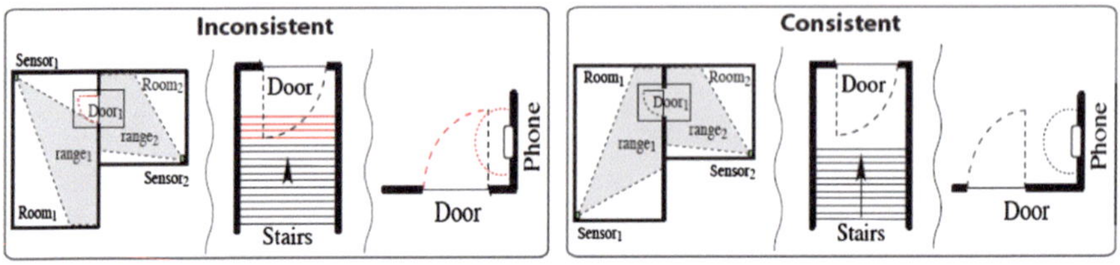

Fig. 6 Design requirements: example spatial interpretations

C1 The FunctionalSpace of the Door of every Office should overlap with the RangeSpace of one or more Camera or MotionSensor.

C2 The StairWay should be topologically non-overlapping with the Functional-Space and OperationalSpace of other entities

C3 People should not be harmed by Doors opening up. In general, the Operation-Space of a Door should be non-interfering, i.e., not overlap with the function/operation (i.e., functional/operational space) of surrounding objects.

The schematization in Fig. 6 is a continuation of the example requirements introduced in (R1–R3), and semantically expressed constraints in (C1–C3). To consider two of the three consistent/inconsistent cases from Fig. 6, namely (C1, C3), below is a semantically grounded semi-formal representation of a requirement constraint:

C1	*Class:*	Qualitative level:DoorFunctionalSpace
	SubClassof:	qualitative level:FunctionalSpace, space:topology:proper part of some (qualitative level:SensorRangeSpace)
C3	*Class:*	qualitative level:PhoneFunctionalSpace
	SubClassof:	qualitative level:FunctionalSpace, not (space:topology:overlaps some (qualitative level:DoorOperationalSpace))

The remaining example from Fig. 6, corresponding to (C2), too may be modelled in a similar manner, namely, as a topological constraint among the primitive conceptual/qualitative/quantitative entities within the design model. Clearly, there are many more possibilities to model requirement constraints on the basis of other aspects of space, e.g., orientation, cardinal directions, metric/fuzzy distances. In this manner of modelling, it must be emphasised that the resulting functional consistency is interpreted strictly with respect to the structural form of the design.

4.4 Design Intelligence—Modes of Inference The term design intelligence is rather open and subject to diverse interpretations; its scope and definition are only limited by the range of the inference patterns that may be operationalised computationally. From the viewpoint of this paper, we have rather specific inclinations with respect to the reasoning capabilities that must be the focus of spatial com- puting for design.

Conceptual Reasoning Conceptual reasoning corresponds to the ontological reasoning patterns that are available within the framework of a terminological reasoning system grounded to the semantics of a Description Logic (DL) [3]. Ontology reasoning systems such as RACER [28], PELLET [43] support typical DL inference tasks at the terminological (subsumption, satisfiability, equivalence, disjointness) and instance levels (instance checking, data consistency, realisation, retrieval). For example:

1. *Retrieval task*: identify all concrete entities/geometric features (e.g., '*polygons*') from instance data coming from a CAAD model that correspond to a design abstraction/artefact such as 'FunctionalSpace' or 'MovableEntity'
2. *Instance checking*: given a set of geometric features within a CAAD model, what is the most general/specific abstract ontological category that the identified feature belongs to from the conceptual/artefactual viewpoint of the designer

From a conceptual reasoning viewpoint, another important reasoning task is determining whether or not the requirement constraints, functional or otherwise, specified by a designer may possibly be satisfied by a model per se. This form of reasoning is useful to check if a given set of design requirement are mutually consistent.

Functional Consistency The example scenarios in Sect. 4.3 illustrated the extent and manner in which functional requirement consistency may be modelled with respect to the structural form of design. This is the form of consistency that has been discussed and il- lustrated throughout this paper. However, the notion of functional consistency transcends beyond the purely spatial aspects of a design, and also includes semi-spatial aspects that include the material and constitution of design artefacts, aspects such as weight, colour, physical characteristics, and artistic aspects that may be beyond the domain of space. Regardless if what precisely what these aspects are, the inference patterns required to ensure functional consistency, in so far as it is formalisable, is essentially some form constraint reasoning approach over a spatial or non-spatial domain, which is the forte of the state-of- the-art in AI research (see Sect. 5).

Hypothetical Reasoning Reasoning about conceptual & functional consistency is only a starting point: for spatial computing, the real challenge of intelligent design assistance is the capability to reason about not what is, but instead about what could be. This form of inference is referred to as hypothetical reasoning. In general, within a decision-support or design assistance tool, metrical changes in the structural layout or changes in the relative spatial relationships of the design elements—i.e., qualitative changes along the conceptual space of the designer—will directly or indirectly entail differing end-product realizations in terms of spatial design requirements, building construction costs, human-factors (e.g., traversability, way-finding complexity), aesthetics aspects, and energy efficiency and long- term maintenance expenses thereof. As such, commonsensical and hypothetical reasoning at the qualitative level about physically realizable[8] and functionally consistent structural forms represents a useful solution approach that is useful for providing the designer with creative design recommendations.

---

[8] Also related is the commonsensical notion of a physically realizable situation defined in terms of physical, compositional and existential consistency of spatial situations [13, 11].

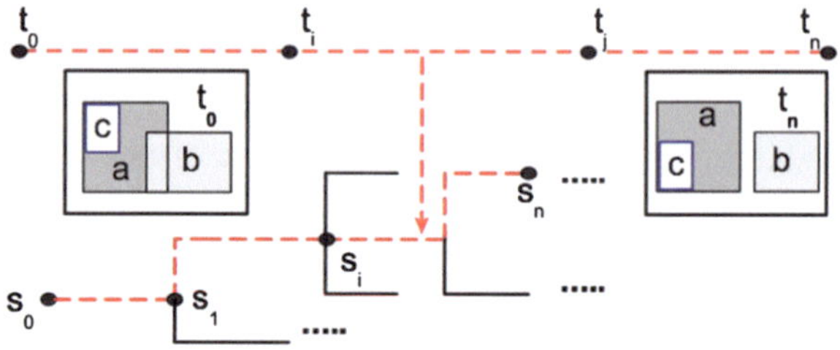

Fig. 7 Branching/Hypothetical situation space

Alternate recommendations are derivable by hypothesizing the possible/potential spatial re-configurations/transformations (e.g., by translation and deformation actions) at the qualitative level; by not discretizing the space and considering the full range of quantitative possibilities, the problem of hypothetical reasoning is in full generality infinitesimal and intractable. As an example, consider the illustration in Fig. 7. The situation-based history <s0, s1,… , sn> represents one path, corresponding to an actual time-line <t0, t1,… , tn>, within the overall branching-tree structured situation space that could be representative of a space of design evolutions at the qualitative level.

Therefore, the objective of hypothetical reasoning about the '*design space*' is to infer/hypothesize (e.g., by abduction) physically plausible qualitative variations in a design that are also essential or functional requirement fulfilling. Indeed, hypothetical reasoning may also take into consideration domain-specific heuristics/physical attributes that determine aspects such as movability, deformability, stability. Such a logic-based approach may also work as a complementary technique to other approaches such as generative and emergent computations.

Assistive Feedback Mechanisms—Design Simulation Assistive feedback mechanisms by visualisation and simulation have to be provided in or- der to communicate diagnostics and other forms of design support within a conventional CAAD workflow. Conventional CAAD tools have remained focussed on providing capabilities for aesthetically appealing 3D visualisation of floor-plans. State-of-the-art tools also allow easy placement/visualisation of third-part 3D models of common interior arte- facts, thereby enhancing the 3D visualisation experience. The human-computer interaction aspects involved in the communication and interaction between the a designer and next- generation CAAD tools is an open topic of research. It is not our objective here to speculate on the future directions of this field of research. The visualisation and simulation aspects pointed out in the following are some benchmarks that have been set for our working pro- totype DSim [12]. DSim attempts to operationalize the concept of being able to "live your design":

- *Semantic browsing* vis-a`-vis the structural hierarchy of the design
- Real-time spatial *artefact simulation* (e.g., sensors, camera; see Fig. 8)• *Inconsistency pinpointing* at the structural and semantic level
- Hierarchical and *selective zooming* for specific requirements
- Automatic *reconfiguration and placement* of design artefacts

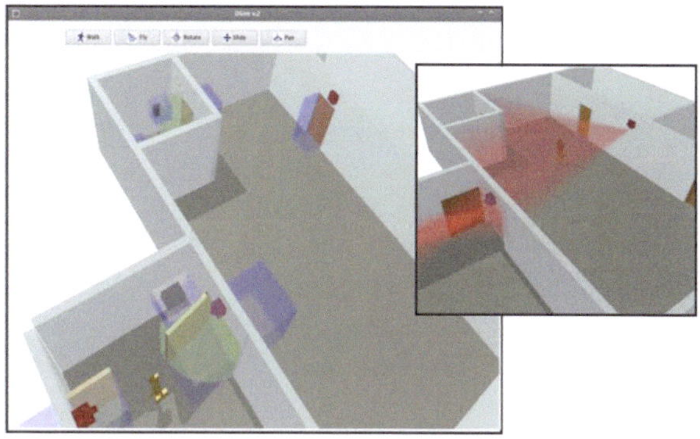

Fig. 8 3D realizations of the *functional*, *operational* and the *range* spaces of the architectural entities. **System DSim**. [12]

We consider the above features to be crucial and necessary for next-generation CAAD tools that not only support the 2D/3D spatial modelling, but also provide the conceptual spatial modelling and functional reasoning capabilities, such as those described in this paper.

Discussion and Summary

We have addressed two themes in this paper: spatial computing for design on the one hand, and the design of spatial computing itself on the other. The main focus here has been on introducing spatial computing for design as a paradigm, the representational and computational aspects that it needs to addresses as a body of work, and finally the concrete application scenarios that it needs to solve. Our notion of spatial computing (for design) is firmly grounded in the AI/KRR-centred perspective, as enshrined in the initial foundations laid out by early pioneers in the field. Gero [24] positioned "*Ten Problems for AI in Design*".[9] With respect to the scope of spatial computing, as addressed in the present paper, we relate to some of them:

- Representation in design, Design semantics—"*What is it that the designer knows and how do we get a computer to know* it?"
- Inference in design—"*much of design inferencing has to do not only with deductive inference but with abductive inference which is concerned with what might be rather than what is*"
- Combinatorial explosion in design—"*as soon as a system deals with what could be it could go on indefinitely*"

The problem of representation in design and design semantics is related to the modelling of multiple-perspectives and the explicit representation of requirements as per their conceptualization by a designer. The problems of reasoning about what

[9] In view of the developments in AI in the last two decades, it is interesting to relate these problems as they existed back then, and as they stand now. We leave this exercise to another paper.

could be and combinatorial explosion in design are two sides of the same coin: hypothetical reasoning (by abduction or otherwise), as positioned in this paper, within a qualitative context, and under additional constraints of physical realizability and architecture domain-specific heuristics is an interesting approach that merits further treatment.

Much has changed in AI since the early 90s. Frame-based systems and semantic net- works have evolved into a range of description logic based ontology languages that are tailored to different levels of expressivity and computational properties [3]. Practical ontology reasoning systems such as *Racer* [28] and Pellat [43] have also come to the fore. The field of qualitative spatial representation and reasoning has emerged has a new discipline within KRR in the last decade—specialized (in- finite domain) spatial reasoning systems *SparQ* [50] and *GQR* [51] now support constraint reasoning and additional application-support services that make it possible to model and reason about spatial knowledge in ways that has not been possible before. Similarly, the evolution of Logic Programming (LP) to Constraint Logic Programming (CLP) [32] and other powerful computational embodiments of the default and non-monotonic reasoning paradigms by way of Answer-Set Programming (ASP) [49] are developments that have only found limited attention in the design community. High-level formalisms to reason about action and change such as the Situation Calculus [36], Event Calculus [33], Fluent Calculus [45], and other more specialized formalisms also similarly grounded in mathematical logic [19], have progressed to the point where prototypical languages (e.g., Indigolog [27], Discrete Event Calculus Reasoner [38], FLUX [27]) allow high-level specification and projective/abductive inference capabilities about dynamic process-like phenomena. These developments open up interesting new possibilities and programming paradigms not only for solving design problems hitherto considered to be computationally intractable, but also for integration, in fundamental ways, of generalised logic-based reasoning on the one hand, and specialized spatial reasoning techniques on the other [10].

The progress made in the last two decades within the knowledge representation and reasoning community in general, and the field of spatial reasoning in specific, warrants a re-visitation into the '*design as problem-solving*' approach of Simon [41]. In spite of garnering initial momentum and interest in the 'AI for Design' community, this approach failed to make an impact by way of practical industrial applications. This paper is partly a statement of our work-in-progress, and partly an attempt to revive some of the basic questions underlying AI in/for design in the context of the specialization we refer to as Spatial Computing.

Acknowledgements We gratefully acknowledge the funding and support by the Alexander von Humboldt Foundation (Germany), and the German Research Foundation (DFG). The paper has immensely benefitted from our discussions and ongoing collaborations with John Bateman, Frank Dylla, Gregory Flanagan, Joana Hois, and Oliver Kutz.

References

1. Akin Ö (1993) Architects' reasoning with structures and functions. Environ Plan B: Plan Des 20(3):273–294
2. Alexander C, Ishikawa S, Silverstein M (1977) A pattern language: towns, buildings, construction. Oxford University Press, New York. ISBN 0195019199
3. Baader F, Calvanese D, McGuinness DL, Nardi D, Patel-Schneider PF (eds) (2003) The description logic handbook: theory, implementation, and applications. Cambridge University Press, Cambridge. ISBN 0-521-78176-0
4. Baldwin C (2007) Steps toward a science of design. In: NSF Principal Investigators Conference on the Science of Design, 2007. http://www.people.hbs.edu/cbaldwin/DR2/Baldwin-ScienceofDesignSteps.pdf
5. Bayazit N (2004) Investigating design: a review of 40 years of design research. Des Issues 20(1)
6. Benedikt ML (1979) To take hold of space: isovists and isovist fields. Environ Plan B: Plan Des 6(1):47–65, January 1979. http://ideas.repec.org/a/pio/envirb/v6y1979ilp47-65.html
7. Bertel S, Freksa C, Vrachliotis G (2004) Aspectualize and conquer. In: Gero J, Tversky B, Knight T (eds) Visual and spatial reasoning in design III. Key Centre of Design Computing and Cognition, Sydney, pp 255-279
8. Bertel S, Freksa C, Vrachliotis G (2007) Aspect-oriented building design: towards computer-aided approaches to solving spatial constraint problems in architecture. In: Allen G (ed) Applied spatial cognition: from research to cognitive technology. Erlbaum, Mahwah
9. Bhatt M (2009) Commonsense inference in dynamic spatial systems: phenomenal and reasoning requirements. In: Bratko I, abkar JŽ (eds) 23rd International Workshop on Qualitative Reasoning (QR 09), Ljubljana, Slovenia, June 2009, pp 1–6, 2009. (Part 1 of 2)
10. Bhatt M (2010a) Reasoning about space, actions and change: a paradigm for applications of spatial reasoning. In: Qualitative spatial representation and Reasoning: trends and future directions. IGI Global, USA, 2010a. http://www.cosy.informatik.uni-bremen.de/staff/bhatt/seer/Bhatt-2010-RSAC-Book.pdf
11. Bhatt M (2010b) Commonsense inference in dynamic spatial systems: epistemological requirements. In: FLAIRS Conference: special track on spatial and temporal reasoning. AAAI Press, 2010b. (Part 2 of 2).
12. Bhatt M, Ichim A, Flanagan G (2010) Dsim: a tool for assisted spatial design. In Proceedings of the 4th International conference on design computing and cognition (DCC'10), 2010. (to appear). BremLBO. Bremische landesbauordnung, 2003. http://www.bauordnungen.de/html/bremen.html
13. Bhatt M, Dylla F, Hois J (2009) Spatio-terminological inference for the design of ambient environments. In: Hornsby KS, Claramunt C, Denis M, Ligozat G (eds) Conference on spatial information theory (COSIT'09). Springer-Verlag, pp 371–391
14. Brown DC (1993) Intelligent computer-aided design (1998 revised version). In Williams JG, Sochats K (eds) Encyclopedia of computer science and technology. Marcel Dekker, New York
15. Brown DC (2007) AI EDAM at 20. Artif Intell Eng Des Anal Manuf 21(1):1–2. ISSN 0890-0604. doi: http://dx.doi.org/10.1017/S0890060407070011
16. Brown DC, Waldron MB, Yoshikawa H (eds) (1992) Intelligent Computer Aided Design, Proceedings of the IFIP WG 5.2 Working Conference on Intelligent Computer Aided Design (Int-CAD91), Columbus, OH, USA, 30 September—3 October 1991, volume B-4 of IFIP Transactions, 1992. North-Holland. ISBN 0-444-81560-0
17. Chandrasekaran B (1990) Design problem solving: a task analysis. AI Mag 11(4):59–71. ISSN 0738-4602
18. Cohn AG. Renz J (2007) Qualitative spatial reasoning. In: van Harmelen F, Lifschitz V, Porter B (eds) Handbook of knowledge representation. Elsevier, Amsterdam

19. Davis E, Morgenstern L (2004) Introduction: progress in formal commonsense reasoning. Artif Intell 153(1–2):1–12. ISSN 0004–3702
20. Dorst K, Vermaas P (2005) John gero's function-behaviour-structure model of designing a critical analysis. Res Eng Des 16(1):17–26, November 2005. doi:10.1007s00163-005-0058-z. http://dx.doi.org/10.1007/s00163-005-0058-z
21. Freksa C (1991) Qualitative spatial reasoning. In: Mark D, Frank A (eds) Cognitive and linguistic aspects of geographic space. Kluwer, Dordrecht, pp 361–372
22. Freksa C (1992) Using orientation information for qualitative spatial reasoning. In: Proceedings of the Intl. Conf. GIS, From Space to Territory: theories and Methods of Spatio-Temporal Reasoning in Geographic Space. Springer-Verlag, Berlin, pp 162–178. ISBN 3-540-55966–3
23. Gero JS (1990) Design prototypes: a knowledge representation schema for design. AI Magazine 11(4):26–36
24. Gero J (1991) Ten problems for AI in design. In: Workshop on AI in Design, IJCAI-91
25. Gero JS (2007) AI EDAM at 20: artificial intelligence in designing. AI EDAM 21(1):17–18
26. Gero JS, Tham KW, Lee HS (1992) Behavior: a link between function and structure in design. In: Brown et al. pp 193–225. ISBN 0-444-81560-0
27. Giacomo GD, Levesque HJ (1999) An incremental interpreter for high-level programs with sensing. In: Levesque HJ, Pirri F (eds) Logical foundation for cognitive agents: contributions in honor of Ray Reiter. Springer, Berlin, pp 86–102
28. Haarslev V, Möller R, Wessel M (2004) Querying the semantic web with Racer + nRQL. In Proceedings of the KI-2004 International Workshop on Applications of Description Logics (ADL'04)
29. Hirtz J, Stone R, McAdams D, Szykman S, Wood K (2002) A functional basis for engineering design: reconciling and evolving previous efforts. Res Eng Design 13(2):65–82
30. Hois J, Bhatt M, Kutz O (2009) Modular Ontologies for Architectural Design. In Proc. of the 4th Workshop on Formal Ontologies Meet Industry, FOMI-09, Vicenza. vol 198 of Frontiers in Artificial Intelligence and Applications. IOS, Italy
31. Horridge M, Patel-Schneider PF (2008) Manchester OWL syntax for OWL 1. 1, 2008. OWL: Experiences and Directioins (OWLED 08 DC), Gaithersberg, Maryland
32. Jaffar J, Maher JM (1994) Constraint logic programming: a survey. J Log Program 19/20:503–581
33. Kowalski R, Sergot M (1986) A logic-based calculus of events. New Gen Comput 4(1):67–95. ISSN 0288–3635
34. Krishnamurti R (2006) Explicit design space?. Artif Intell Eng Des Anal Manuf 20(2):95–103. ISSN 0890-0604. doi: http://dx.doi.org/10.1017/S0890060406060082
35. Loos A (1930) Ornament and crime. Reprint Vienna, Innsbruck
36. McCarthy J, Hayes PJ (1969) Some philosophical problems from the standpoint of artificial intelligence. In: Meltzer B, Michie D (eds) Machine intelligence 4. Edinburgh University Press, Edinburgh, pp 463–502
37. Moratz R (2006) Representing relative direction as a binary relation of oriented points. In: ECAI, pp 407–411
38. Mueller ET (2006) Commonsense reasoning. Morgan Kaufmann Inc., San Francisco. ISBN 0123693888
39. Randell DA, Cui Z, Cohn A (1992) A spatial logic based on regions and connection. In KR'92. Principles of knowledge representation and reasoning. Morgan Kaufmann, San Mateo, pp 165–176
40. Renz J, Nebel B (2007) Qualitative spatial reasoning and using constraint calculi. In: Handbook of spatial logics. Springer, Dordrecht, pp 161–215
41. Simon H (1969) The sciences of the artificial [The Karl Taylor Compton lectures.]. Chapter 3: the science of design. MIT, Cambridge
42. Simon H (1996) The sciences of the artificial, 3rd edn. MIT, Cambridge, ISBN 0-262-69191–69194

43. Sirin E, Parsia B, Grau B, Kalyanpur A, Katz Y (2007) Pellet: a practical owl-dl reasoner. Web Semantics: Science, Services and Agents on the World Wide Web, 5(2):51–53, June 2007. ISSN 15708268. doi:10.1016/j.websem.2007.03.004. http://dx.doi.org/10.1016/j.websem.2007.03.004
44. Sullivan L (1896) The tall office building artistically considered. Lippincott's Magazine
45. Thielscher M (1998) Introduction to the fluent calculus. Electron Trans Artif Intell 2:179–192
46. Thielscher M (2005) Flux: a logic programming method for reasoning agents. Theory Pract Log Program 5(4–5):533–565. ISSN 1471–0684
47. US GSA. US Courts Design Guide, 2007. URL http://www.gsa.gov/graphics/pbs/Courts_Design_Guide_07.pdf. Judicial Conference of the United States. US General Services Administration (GSA). April 23 2010
48. Vermaas P, Kroes P, Light A, Moore S (eds) (2008) Philosophy and design: from engineering to architecture. Springer, Dordrecht
49. Vos MD (2009) ASP: the future is bright. In: LPNMR, pp 625–627
50. Wallgrün JO, Frommberger L, Wolter D, Dylla F, Freksa C (2007) Qualitative spatial representation and reasoning in the sparq-toolbox. In: Barkowsky T, Knauff M, Ligozat G, Montello D (eds) Spatial Cognition V: Reasoning, Action, Interaction: International Conference Spatial Cognition 2006, vol. 4387 of LNCS, pp 39–58. Springer-Verlag Berlin Heidelberg
51. Westphal M, Woelfl S, Gantner Z (2009) GQR: A fast solver for binary qualitative constraint networks. In: AAAI Spring symposium on benchmarking of qualitative spatial and temporal reasoning systems

SIRN—Synergetic Inter-Representation Networks: An Approach to Urban Planning and Design with Implications to Visual Reasoning and Design Creativity

Juval Portugali

The term SIRN integrates two notions: *IRN—Inter-representation network* [1] and *synergetics* which is the name assigned by Haken to his theory of complex, self-organizing systems [2]. *SIRN* was originally developed by Haken and Portugali as an approach to, and model of, cognition and cognitive mapping [3]. The aim of this paper is twofold: firstly, to relate SIRN to cognitive planning and urban design and secondly, to discuss the implications thereof to visual-spatial reasoning and design creativity. The discussion below starts with an overview on planning design and construction. It then introduces the notion and models of SIRN. Next a SIRN view on planning and design is suggested and finally on visual-spatial reasoning and design creativity.

Planning, Design and Construction

In the domain of cities, the production of artifacts (buildings, road nets, neighborhoods, cities …) commonly takes three forms: *planning*, *design* and *construction*. While all three are processes of production, they differ in the nature of their end product: the product of planning is a *plan* such as a land-use plan or a set of policies about a given area; the product of design is some *model* of the end product, such as a graphical *sketch* of it or, 2D and 3D drawings, or a 3D physical model, or a computerized VR of it; the product of construction is the end product itself, e.g. an urban neighborhood. Obviously, the three are not independent of each other: design always involves planning while planning might involve design (e.g. land-use map) but not always (e.g. when it ends with a set of policies). In a similar way, construction involves planning and design while the latter two often involve construction of a sort, but usually not of the final product.

J. Portugali (✉)
Tel Aviv University, Tel Aviv, Israel
e-mail: juval@post.tau.ac.il

J. S. Gero (ed.), *Studying Visual and Spatial Reasoning for Design Creativity*,
DOI 10.1007/978-94-017-9297-4_8

Fig. 1 The tower of Hanoi game

Of the three, planning and design are commonly regarded as 'cognitive' and are thus associated with specific research domains known as *cognitive planning* [4] and *design thinking* (*or, design cognition*) [5]. Cognitive planning as developed by psychologists and cognitive scientists commences with the notion that planning—that is, the ability to think ahead to the future and to act ahead toward the future—is also a basic cognitive capability of humans.

Design cognition as developed by architects, computer scientists and others, commences with the notion that the process of design is 'cognitive' due to the fact that it is associated with a whole set of general cognitive capabilities such as thinking, imaging, planning and the like, and specific cognitive capabilities such as visual thinking and spatial reasoning. To the latter one might add projects such as Hillier's [6] and Alexander's [7] that see design in terms of a two-ways interaction between the designed world of artifacts and the human mind and behavior: On the one hand, the process of design involves several cognitive processes (such as imaging, thinking, representation and so on) while on the other, the world of artifacts affects human cognition and behavior.

From the perspective of cognitive planning a plan might be a kind of *prospective memory* task: e.g. 'to buy flowers on the way home back from work' [8]. However, a plan might also be a sequence of actions that lead to a desired final state such as in *The Tower of Hanoi* game (Fig. 1). In the tower of Hanoi the final state is definite and pre–determined. There are, however, tasks in which the desired final state is vague or not known at all at the beginning of the process. For example when an architect starts to plan a house or an engineer starts to plan a machine that doesn't yet exist. These latter cases take us to the distinction between *well-defined* versus, *ill-defined problems that* require creative solutions (see below).

Implicit in both cognitive planning and design cognition is the view that construction—the third step in the process of production—is not cognitive. As exceptions to this view one can mention Hillier with his space syntax [6], Alexander and his studies ranging from the *Pattern Language* to *The Nature of Order* [7] and more recently Nili Portugali with her *Holistic Approach to Architecture* [9]. SIRN belongs to these exceptional views.

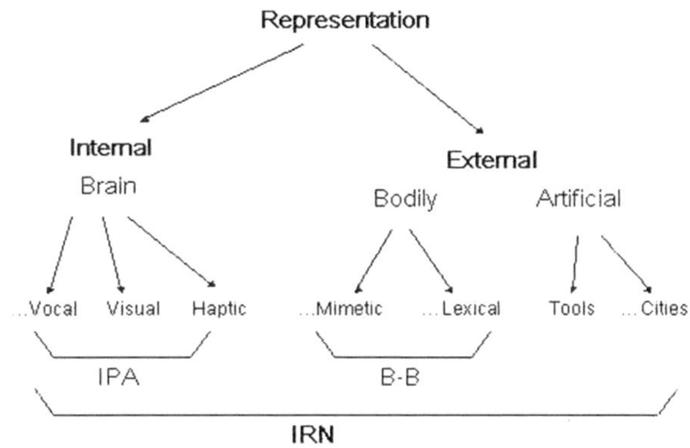

Fig. 2 Internal and external representation

SIRN (Synergetic Inter-Representation Networks)

This section first introduces IRN and synergetics—the two components of which SIRN is composed—then, the relations between them and the way SIRN is related to *classical* and *embodied* cognition—the two major paradigms that currently dominate cognitive science—and finally to the way artifacts as external representations convey information.

IRN—Inter-Representation Networks

According to IRN [1, 10], the cognitive system in general, and the one associated with cognitive maps in particular, extend beyond the individual's mind/brain into the external environment. This is so in the sense that the cognitive system is a network composed of internal and external representations. Internal representations refer to entities constructed by the brain that represent information of the external environment, while external representations to entities constructed by means of humans' mimetic, linguistic and artifact-making capabilities that represent information generated by the mind/brain.

More specifically, the claim is twofold: Firstly, that *humans have an innate capability for representation that comes in two forms: external and internal*. This is illustrated in Fig. 2. Internal representations are the outcome of brain processes the end product of which is various forms of information—visual, olfactory, haptic, lingual, etc., combination thereof, as well as emotions, intentions and the like—that are enfolded (i.e. represented) in the matter of the brain. External representations refer to behavior, actions and products that represent internal representations. External representations can be further divided into *bodily* and *artifactual* representations. Bodily representations are representations made by the body and never extend beyond it, such as mimetic or lexical representations. Artifactual representations are made by the body, but extend beyond it to become stand-alone artifacts.

The notion of internal representation as utilized here is close to what Varela et al. have termed "weak representation"—in contrast to "strong representation" that typifies classical cognitivism [11]. According to this view, the brain generates patterns or internal representations in the form of images, cognitive maps and the like. However, such patterns "are not stored in any static way, neither with respect to geographical areas, nor with respect to mode of representation. They are dynamically created anew, each time, as ad hoc entities: The brain is capable of creating a multiplicity of cognitive maps with specific perspectives, scales and modes, by means of learned synaptic connection strengths that govern the cooperation between the neurons" [1, p. 6], [10]. Internal cognitive processes thus involve an interaction between symbolic internal representations, but not as static, fixed, stored, internal representations, as implied by the computer metaphor to cognition. The same holds true for bodily external representations. They are also ad hoc entities, created each time anew, when we mimetically or lexically externalize some of our emotions, ideas and other internal representations. *The case is different, however, with artifactual external representations, that is, with what has been described above as the products of planning (a plan), design (a model) and construction (the end product itself). They do enfold information or store symbols in static ways. Tools, texts, plans, models, neighborhoods and whole cities are typical examples in this context.*

The second claim is that many cognitive processes, planning, design and construction included, evolve as an interaction between internal and external representations. The major source of inspiration for this proposition has been Bartlett's (1932/1961) scenarios of serial reproduction devised by him in his study on Remembering. A typical Bartlett scenario evolves like this (Fig. 3): a test person is given a text or shown a figure and is asked to memorize it. He or she is then asked to externally reproduce it out of memory, by rewriting the text or re-drawing the figure. This externally represented text or figure is given to another test person and so on. The usual result of such scenarios is that after several strong fluctuations in the reproduction, the text or the figure stabilize and do not change much from iteration to iteration. Bartlett reports that the same happens when the experiments are carried out with a single person.

Bartlett developed the method of serial reproduction as a means to externalize and expose the otherwise internal processes and representations and as means to develop the notion of schemata. From the point of view of SIRN, the interpretation is that the process exemplifies an emergent, task-specific, cognitive network of internal and external representations, and a sequential interplay between them [1, 3, 10]. To see how, we have to look at the second component of SIRN, namely, at Synergetics.

Synergetics

Synergetics is the name Haken assigned to his theory of complex self-organizing systems [2]. The theory is regarded as one of the foundations of a growing body of studies that currently is grouped under the title of *complexity theory* [13, 14].

Fig. 3 Chronological interactions of internal and external representations (after Bartlett)

Originating in the 1960s in physics, in connection with phenomena such as laser and fluid dynamics, synergetics has become a general paradigm that is intensively applied to several domains including cognition, brain functioning and brain dynamics and also to cities and planning [13, 14].

The theory deals with systems that are *open*, in the sense that they exchange mater, energy and information with their environment and *complex* in two respects: First, in the sense that their parts are so numerous that there is no technical way to establish causal relations among them. Second, their parts and components are interconnected in a nonlinear fashion by a complex network of feedback and feed-forward loops. A central property of such a system is that its action and behavior are not determined causally by means of forces that are acting on it, but rather internally and spontaneously by means of its internal dynamics. The forces acting on such systems trigger internally spontaneous dynamics, which then determine the system's structure and behavior. Such systems thus self organize their structure, action, behavior or output—hence the notion of *self-organization* which is a fundamental property of open and complex systems that attain their order spontaneously and are further typified by phenomena of non-causality, nonlinearity, instability, fractal structure and chaos.

Synergetics—the working together of many parts—is the central property of Haken's theory of complexity. Its second property and methodological guide, is to "look for qualitative changes at macroscopic scales" [15, p. 39]. The theory thus focuses on the working together of many parts in relation to qualitative changes at systems' macroscopic scales. Synergetics was developed by reference to specific case studies that became its basic paradigms: The *laser* paradigm, the *fluid dynamic* paradigm, the *pattern recognition* paradigm, and the *finger-movement* paradigm. The scenario common to all the various cases can be described as follows: A given internal or external *control parameter* that is acting on the system triggers a chaotic movement and interaction between its many parts. This chaotic movement is interpreted as a situation by which several systemic order states compete among themselves. When the control parameter crosses a certain threshold, the hitherto chaotic form of movement and interaction suddenly and spontaneously gives rise to a coherent movement and interaction where all the parts behave in a concert. This coherent movement is termed *order parameter,* and the process by which the many parts abruptly "obey" the order parameter and in this way support and reproduce it—the *slaving principle.*

A central effort of Haken and co-workers' research in the last three decades was in the domain of cognition and brain functioning [15] as noted. The basic proposition in these applications is that the brain and its various cognitive systems are open, complex and, therefore, self-organizing systems and that their dynamics is best described by the principles of synergetics. A case in point is pattern recognition (Fig. 4): the cognitive system (brain or computer) is given a few features of a certain pattern (i.e. a face or a city map) referring to one out of a repertoire of patterns which are stored in the brain/computer. This triggers a process of self-organization in which several order-states emerge and enter into a competition. This competition is resolved when a certain order parameter "wins", enslaving the various features by means of associative memory, and a recognition is established.

With respect to behavior, synergetics suggests seeing brain, cognitive and behavioral processes in terms of open, complex, task-specific/context-dependent systems that achieve their coherence spontaneously, by means of a complex co-operation and interaction between their huge numbers of parts. For example, 'talking', or 'speech production', is a behavior and action that requires the emergence of a task-specific complex system that includes brain neurons, muscles, joints, and so on, and a specific synergy between these parts. The interacting elements of that system are, therefore, both internal and external.

To this view SIRN adds, first, that in some tasks the synergy between the many internal and external parts of the system gives rise to internal and external representations. Second, that in such cases the process often evolves as an interaction between internal and external representations. Third, that in many tasks, such as reading, painting, sculpturing, writing, discourse, carpentry, architectural and urban design, pottery making, navigating and/or shopping in a city, the parts of the task specific-system include, in addition to neurons, muscles and joints, stand-alone artifacts.

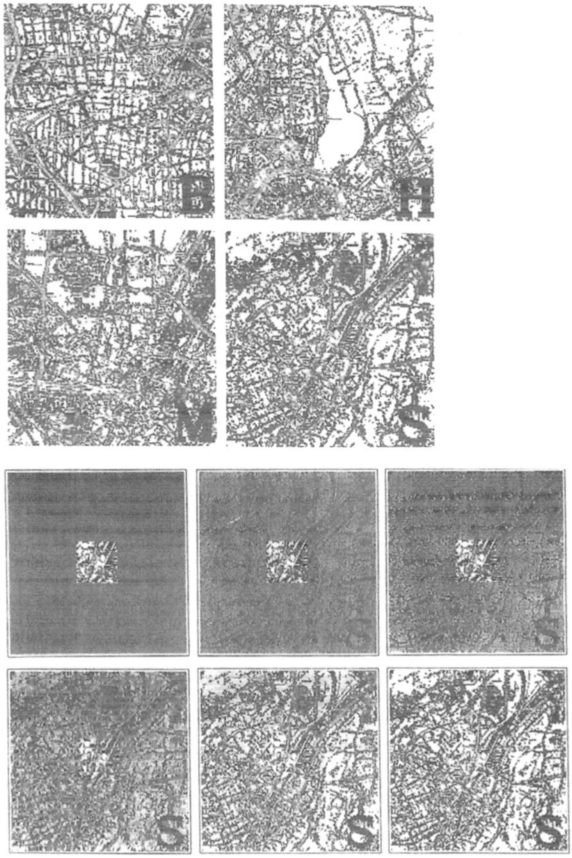

Fig. 4 Haken's pattern recognition

From IRN to SIRN

Synergetics thus adds to the notion of IRN the view that the brain/mind, cognition, cognitive mapping, and the interaction between internal and external representations, are all complex self-organizing systems that evolve in line with the principles of synergetics. SIRN is thus a model and a theory that casts IRN into the formalism of synergetics [3].

The transition from IRN to SIRN can be illustrated by means of a distinction between cognitively *simple*, *complicated* and *complex* tasks [14]. *Cognitively simple tasks* are tasks that can be performed by working memory (e.g. $2\times 3=6$) while *cognitively complicated tasks* (e.g. $257\times 389=99973$) are tasks that cannot be performed by working memory due to the "magic number seven plus minus two" that constraints our ability to process information in working memory [16]. One way to overcome this limitation is by means of IRN: We first externalize the task (write it down on a paper); then we solve part of it internally ($8\times 7=63$); externalize it again and so on in a sequence until the task is completed.

Cognitively complex tasks refer to *creative* cognitive tasks, when a person writes, paints, designs etc. Such a task often starts with a vague idea in mind that the person then externalizes by writing it down or by painting it, for instance. Here too the process proceeds by interplay between internal and external representations, but with one important addition—*it involves emerging properties*. It is here where synergetics (and complexity theory in general) gets in and the process becomes SIRN. More specifically, the process might start with a preliminary internal idea (or external cue that entails internal idea) that the person then externalizes and so on. After a few internal-external iterations an order parameter (in the sense of synergetics) emerges and enslaves subsequent iterations.

The distinction between cognitively complex tasks and the cognitively simple and complicated tasks is similar to the distinction between *well-* and *ill-defined problems* that is common in the domains of planning and design. Simple and complicated cognitive tasks are two aspects of well-defined problem, namely, problems that can be solved by a single cognitive act (e.g. $2 \times 4 = 8$) and problems that can be solved by an algorithm, namely, by a sequence of cognitive steps that necessarily leads to the required solution (e.g. $257 \times 389 = 99{,}973$). On the other hand, cognitively complex tasks correspond to ill-defined problems in the sense that their solution requires a sequential stepwise SIRN play between internal and external representations, some of which are creative, that is, novel steps and solutions that are not known at the start of the process. However, while creative solutions are not known at the start of the process, experience indicates that they often emerge as the SIRN process advances.

The Paradigmatic Context

Since its emergence in the mid 1950s, cognitive science—*The Mind's New Science* as Gardner called it [18]—evolved under two main paradigms: *classical cognitivism*, or the *information processing approach* (*IPA*) that dominated the field in its first decades and *embodied cognition* that was introduced in the 1970s. Classical cognitivism assumes a complete separation between brain/mind, on the one hand, and bodily action, on the other. As illustrated in Fig. 5A, cognitive processes such as perception are treated as conceptually separate from bodily action. More recently, we see a shift toward the pragmatist embodied cognition approaches, according to which bodily action is part of the cognitive system. Hence the notion of *Perception–Action* (Fig. 5B).

SIRN can be seen as an extension of the action-perception view. It starts from the observation that the new developments that emphasize *embodied* cognition, *action*-perception and task-specificity, imply also the "legitimization" of artifacts. Bodily artifacts such as talking, grabbing, walking etc., are regarded by these approaches as integral elements of the cognitive process itself. Within this context, Gibson has shown that, in some tasks, stand-alone artifacts such as tools function as an extension to the body [19, 40, Fig. 1]. SIRN adds to the latter that in certain tasks

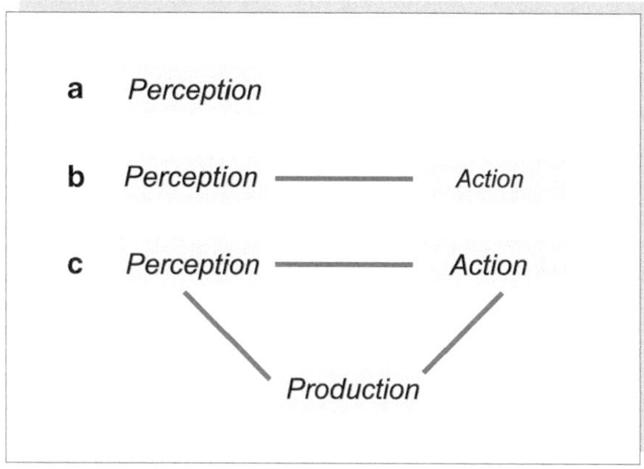

Fig. 5 Gardner's summarization of the progression of cognitive science

and contexts, stand-alone artifacts and the process of their production function not only as an extension to the body, but also as an extension to the mind. In the latter cases, the cognitive process and system includes perception, action and *production* (Fig. 5c).

SIRN thus suggests perceiving the cognitive system as a network composed of individual *and* collective cognitive elements. These elements are termed *representations*. They may be ad hoc products of neural activities in the brain, but also products of bodily activities in the environment. In the first case they are termed *internal representations*, in the second, *external representations*. A key feature of SIRN is that the various representations form a system and a network—an *Inter-Representation Network (IRN)*. Another element is that the parts of that system—its internal and external representations—do not pre-exist as atomistic entities. Rather, they are ad-hoc entities that emerge out of the dynamics. Cognition, therefore, is not a manipulation of stored internal representations, as implied by the computer metaphor of classical cognitivism, but a dynamic process that gives rise to various representations that, by means of their interaction, give rise to a task-specific/context-dependent cognitive system. In some tasks and situations the emergent system is composed of internal representations, in others of internal *and* external representations. In some tasks and situations the latter might be mimetic or lexical behaviors, in others stand-alone artifacts. These task-specific/context-dependent systems attain order, become relatively autonomous, and as such interact with other systems and with the environment. All this happens spontaneously by means of self-organization.

Shannonian and Semantic Information

The above discussion entails a question: In what sense externally represented artifacts are cognitive? The answer: in the sense that they are information carriers.

In their paper "The face of the city is its information", Haken and Portugali demonstrate that artifacts *are entities that enfold, convey and can thus be cognised in terms of, two forms of information: Shannonian and semantic* [20]. Shannonian information is information as defined by Shannon's theory of information [21]. That is, a quantity (usually measured by *bits*) that indicates the information capacity of a communication channel, irrespective of the quality or meaning of that information. It is "information with meaning exorcised" [22]. Subsequent studies have shown that the notion 'communication channel' might refer not only to literal channels (i.e. telephone), but also to humans' short-term memory [16], Gestalt shapes [23] and even urban forms such as buildings, streets or whole cities [20, 24]. Semantic information, on the other hand, refers to the *meaning* conveyed by a representation as perceived by a specific receiver. The process of pattern recognition, of a face or a cityscape, is a typical example here: one sees a shape (a face, a prominent landmark) as well as the meaning it conveys (a certain person in a good mood, the tower of a holy Gothic cathedral, etc.).

Semantic information can further be divided into *innate semantic information* and *experiential semantic information.* I*nnate semantic information* refers to the properties of representations as perceived by animals and humans by virtue of their innate cognitive capabilities. A tower at the center of a flat monotonous city, for example, will be cognised as a landmark by every human individual because of its geometry alone. This kind of information is related to Gibson's ecological approach and his notion of *affordance* [19]. It suggests that objects afford information by virtue of the specific relations that emerge from the interaction between their physical properties and the psychophysical properties of the perceiver. Such information applies to every individual.

Experiential semantic information, on the other hand, refers to the meaning of representations as perceived by humans by virtue of their subjective personal and cultural (inter-subjective) experience. The above noted tower at the center of a flat monotonous city can again serve as an example. Here, however, due to its shape and its steal and glass structure, the tower indeed represents a landmark, but also technological sophistication, economic power, and so on. The tower thus enfolds culturally specific *meaning.*

SIRN—The Basic Model and its Submodels

In a paper from 1996 Haken and Portugali have cast the notion of SIRN into the formalism of synergetics [3]. This was done, by developing a general SIRN model and by deriving from it three submodels that refer to the way an interacting network of internal and external representations is related to the cognition and (spatial) behavior of, firstly, a single person (the *intra-personal submodel*), secondly, several persons acting sequentially (the *inter-personal submodel*) and finally, many persons acting simultaneously (the *interpersonal with a common reservoir submodel*). In this section I present the graphical part of these models.

Fig. 6 Internal and external representations

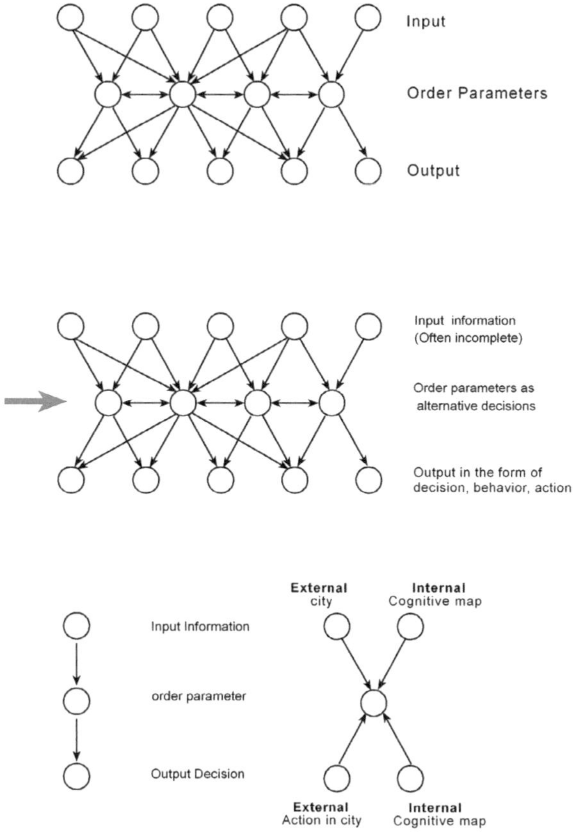

These three submodels are based on the observations that while many cognitive capabilities are essentially solitary, personal and subjective, several cognitive capabilities are by their nature collective. The notion of "brain storming", for example, refers to the fact that a lot of thinking takes place collectively. The same applies to planning, design and construction that stand at the core of our present discussion. People tend to plan, design and produce artifacts collectively, at a variety of groups' form and size, ranging from friends, families and firms to professional planners or designers in commercial, national and urban planning teams. In a recent paper I've suggested referring to such planning as *collective planning* [14].

The Basic SIRN Model

The derivation of the basic SIRN model starts with Haken's (1991/2004) 'synergetic computer' as presented in Fig. 6 (*top*). This is a fully parallel computer that represents an alternative to the conventional neural network model in that the elements of

Fig. 7 The SIRN model

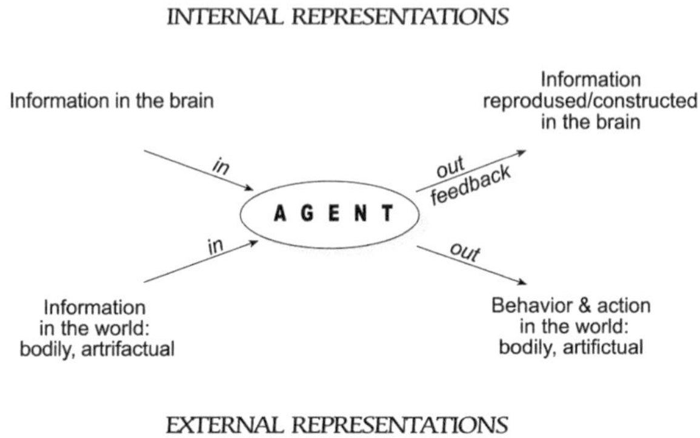

its inner layer are order parameters. As can be seen, it is composed of an input layer with model neurons representing the initially given input activity; a middle layer representing the order parameters, and an output layer with neurons representing the final activity of each neuron. The first step is to look at this network from the side, as indicated by the arrow in Fig. 6 (*middle*). The result is shown in Fig. 6 (*bottom, left*). Adding to the latter external inputs and outputs, we arrive at our basic SIRN model in Fig. 6 (*bottom, right*). As can be seen, it has two kinds of inputs, internal and external and two kinds of outputs, again internal and external. The middle node symbolizes the order parameters that emerge out of the dynamic interaction between internal and external representations [3]. The basic SIRN model as derived from Fig. 6 is illustrated graphically in Fig. 7.

Figure 7 can be seen as symbolizing a self-organizing active agent that is subject to two flows of information: internal and external. The first is coming from the mind/brain, in the form of ideas, fantasies, dreams, thoughts, and the like, while the second from the 'world'—via the senses, the agent's body and/or artifacts. The interaction between these two flows gives rise to an order parameter that governs the agent's action and behavior, as well as the feedback information flow to the agent's mind. 'Action or behavior' may refer (see Proposition two) to a single individual executing exploratory behavior, reproducing texts or drawings in the Bartlett scenarios, as well as to several individuals collectively reproducing large-scale artifacts. In an analogous fashion, the 'feedback information flow' refers to the formation of internal representations, such as images or learned patterns. The order parameters are determined by a competition in line with the synergetics' pattern recognition paradigm noted above. It is important to note that all the steps indicated above (and in the sub-models below), can and have been, performed by a computer so that the approach is entirely operational.

Intrapersonal Subjective Submodel

Consider a person writing a paper, or a story, or a person painting or sculpturing, or designing, or building a certain structure. S/he probably starts by externalizing a

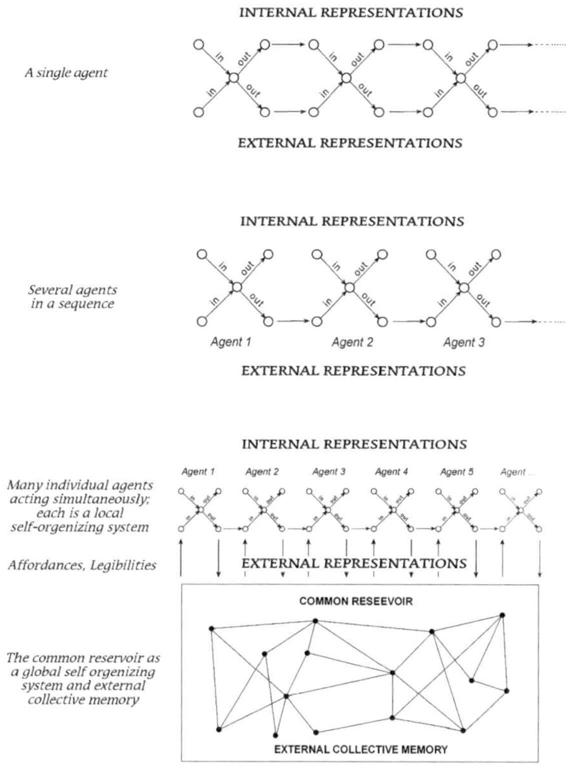

Fig. 8 Sequential iterative interaction between internal and external representations.

given vague idea s/he had internally represented in mind. This first external representation feed back to the person's mind and re-shapes his/her internal representation and so on in a sequential iterative interaction between internal and external representation as graphically described by Fig. 8 (*top*). According to synergetics after a few such iterations an order parameter emerges and enslaves the cognitive system and brings it into a steady state.

As described by past SIRN studies such processes are typical of animal behavior as well as of human behavior—see for example, [10, 14] and further examples and bibliography there. For the present discussion consider, first, Brancusi's Kiss (Fig. 9). As discussed in a previous study [10] this work of Brancusi evolved as a typical process of serial reproduction with several similarities to the Bartlett's scenarios. One can see how the form of a single object/product is changing in time by means of a SIRN internal-external dynamics and how it becomes more and more abstract.

A second example is Picasso's Guernica (Figs. 10, 11) that has been studied by several scholars as means to illustrate the production of artwork and the role of sketches in the process—Arenheim's *Picasso's Guernica* is an often quoted study. The Guernica differs from the Kiss in the time scale—the Kiss evolved slowly from 1907 to 1937, while the Guernica in a short time during the year 1937, after the

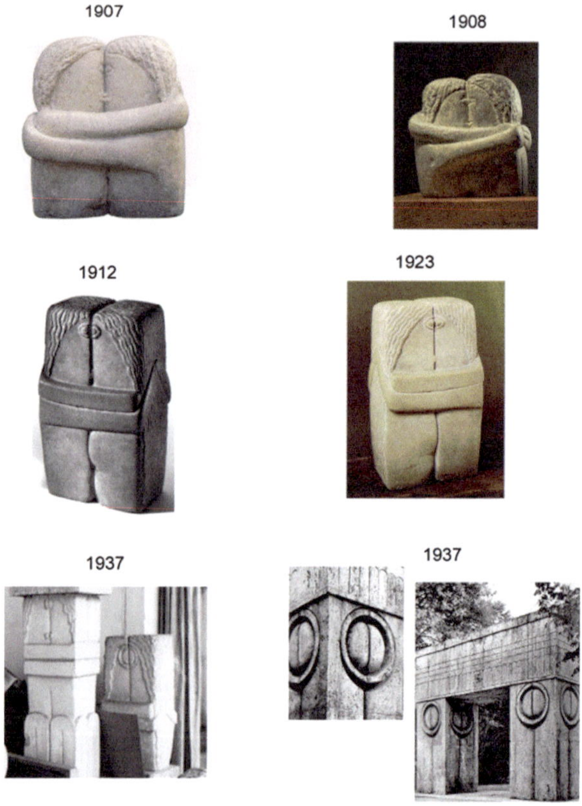

Fig. 9 Brancusiv

town Guernica was bombed. In the case of the Guernica we are exposed also to the preparatory work—the partial sketches and the painting itself in its various stages of evolution/development.

The role of sketches in architectural design and its relation to visual thinking has been subject to several studies by Gabi Goldschmidt [26, 27].

Interpersonal Collective Submodel

This is the classical Bartlett scenario, as illustrated above in Fig. 3. A typical experiment starts, as we have seen, with a given external input and proceeds with a sequence by which each person's externalized reproduction of the remembered input becomes an input to the next person to remember and externalize, and so on. As in the intrapersonal case, after several initial steps that exhibit major changes from one reproduction to the other, the story or the drawn figure stabilizes and does not change significantly from iteration to iteration. In the language of synergetics we would assert that a certain order parameter has enslaved the system and brought it to a steady state.

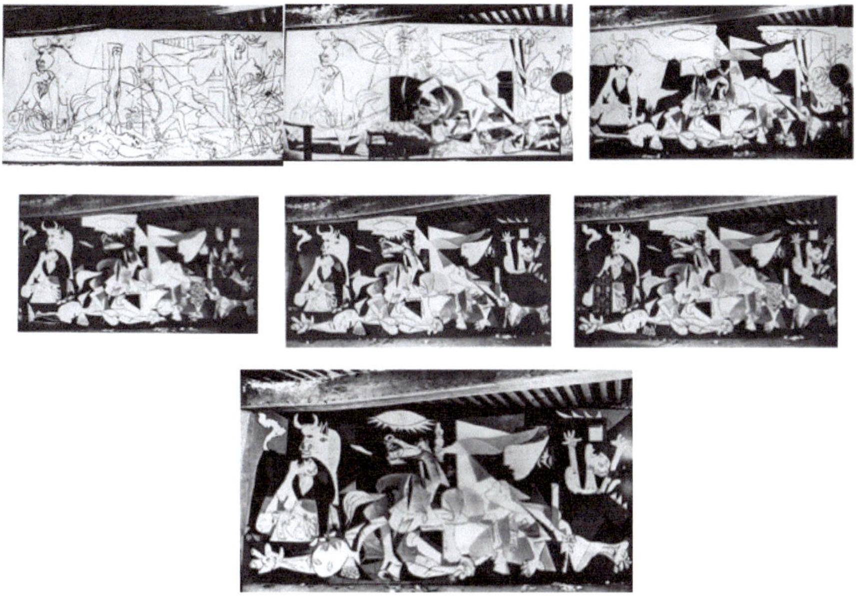

Fig. 10 Guernica—the sequence

Fig. 11 Guennica—the sketches

Fig. 12 Development of a lamp through sequential interactions of external and internal representations

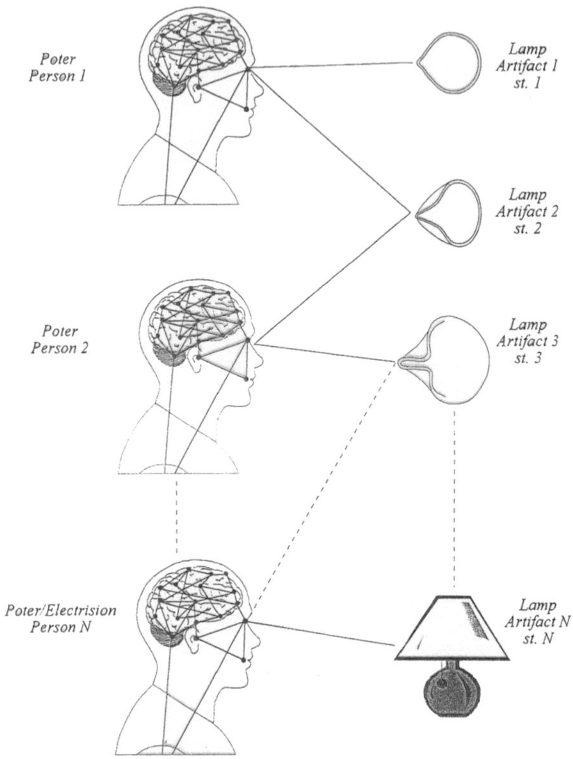

This interpersonal process implies that several persons, with their individual-subjective cognitive systems, participate in producing an externalized collective cognitive product, without being aware of their collective enterprise. As this sequential process evolves, and its collective product constructed, each individual's externally represented reproduction gradually becomes "more" collective and so does each individual's internally represented remembering. The individuals engaged in the process are thus being 'enslaved' by the collective order parameter that emerges in the process. Figure 8 (*middle*) graphically describes this interpersonal scenario by means of our SIRN model.

An interesting example of this SIRN process is the evolution of ceramic artifacts. Consider for this purpose Fig. 12 (*top right*) that shows an oil lamp from the Middle Bronze period (some 3,750 ago). Nowadays, however, the term "lamp" refers mainly to electric lamps, but we still have oil lamps here and there. The archaeological record teaches us that, similarly to many other artifacts, the evolution of lamps is typified by morphological continuity and self-similarity: lamps that are morphologically similar to each other tend to be close in time and space. This property, which is specifically prominent with ceramic artifacts, provided the basis to Sir F. Petrie's notion of *relative chronology* [28].

Fig. 13 The city game

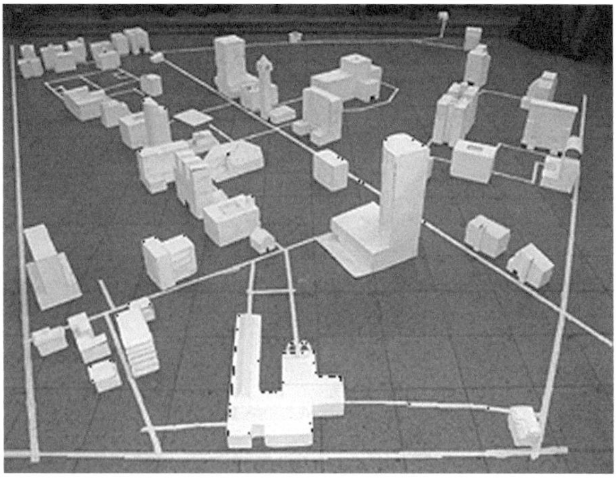

Figure 12 graphically illustrates the process of a lamps' production. It starts with a certain Lamp 1 that already exists in the world. Potter 1 looks at that lamp and internalizes its form in his/her memory. Potter 1's memory (mind/brain) then sends orders to his/her hands that then produce Lamp 2 and so on. Note, firstly, that lamps 1 and 2 are mediated by Potter 1's mind/body, lamps 2 and 3 by Potter 2's mind/body and so on. Secondly, that due to the fact that each pair in the sequence of lamps is mediated by a potter's mind/body, there is always a possibility for a copying mistake—to what Cavalli-Sforza and Feldman have termed "cultural mutation" [29]. Thirdly, that we have here a play between ad-hoc internal and external representations that emerge in the production process: The externally represented Lamp 1 becomes an internal representation in Potter 1's mind and the basis for Potter 1's external representation in the form of Lamp 2 that then becomes an internal representation in Potter 2's mind and so on in a play between internal and external representations.

Interpersonal with a Common Reservoir Submodel

This submodel suggests that the SIRN processes participate also in the emergence of large-scale collective artifacts such as cities. To illustrate and study such processes Portugali devised a set of experiments—*city games*—that can be viewed as a new type of the above Bartlett scenarios [1]. Their essence is a process of sequential reproduction that is interpersonal, collective, and in addition *public*—the participants observe the game as it develops. Each player in the game is given a 1:100 mockup of a building, and in his/her turn is asked to locate it in the virtual city on the floor. In a typical game (Fig. 13), the players observe the city as it develops, and in the process also learn the spontaneously emerging order on the ground. After several initial iterations a certain urban order emerges. The participants internalize this emerging order and tend to locate their buildings in line with it. Such an experiment includes all the ingredients of the SIRN process: a sequential interplay between internal and external representations, the emergence of a collective complex city as an artifact, and a typical synergetic process of self-organization as demonstrated below.

Figure 8 (*bottom*) illustrates graphically this public-collective SIRN sub-model. Each individual player/agent is subject to internal input constructed by the mind/brain, and external input which is the legible information coming from the common reservoir. In the above city game it is the virtual city on the ground; in real life, it is the real city. The interaction between these two forms of input gives rise to a competition between alternative decision rules that ends up when one or a few decision rules "wins". The winning rule(s) is/are the order parameter(s) that enslave(s) the system. The emerging order parameter governs an external output, which in the city game is the player's location action in the city, and an internal output, which is an information feedback loop back to the mind/brain.

Both the previous sub-model and the present one involve a dual, two-scale, self-organization process: an individual-local scale referring to each individual agent as a self-organizing system, and a collective-global scale, referring to the whole city as a self-organizing system. The individual agents by their action and behavior determine the city, which by means of its emerging order parameter(s) enslaves the minds of the individual agents. In the language of synergetics this process is termed *circular causality*. In terms of social theory it is close to notions of socio-spatial *reproduction* and *structuration*. As illustrated by Portugali, the common reservoir might be a non-biological externalized memory such as a city, a planning textual report or an urban planning policy emerging out of a discourse among the members of a planning team [13]. Note that as in the previous model, here too, due to circular causality, as the process evolves the subjective cognitive maps of the individual agents are becoming more similar to each other and an inter-subjective, collective cognitive map emerges. Both private–subjective cognitive maps and public-collective ones are thus *constructions*.

SIRN Perspective on Planning and Design

Since its first introduction in 1996 the notion of SIRN was further developed and applied to several research domains, mainly to urban dynamics and planning while recently some preliminary steps have been made to apply SIRN to urban design. These applications are described below.

Decision Making

Haken and Portugali have applied the above SIRN models to decision making in the context of planning [13]. They started from an analogy between decision making in the context of planning and design and pattern recognition as a basic cognitive capability of humans. It is typical in pattern recognition that an organism is exposed to some partial information on the basis of which, by means of associative memory, it then recognizes the whole pattern. This cognitive task has been a focal issue in complexity theory in general and synergetics in particular. Figure 4 above illustrates a typical case.

The analogy to planning and design, according to Haken and Portugali is the following: as in pattern recognition, so in decision making, the process starts with partial information about the final pattern and/or planned object; and as in pattern recognition so in decision making, the brain/mind/computer complement the needed information by means of associative memory in the form of analogies, metaphors and the like. Furthermore, as in pattern recognition so in planning/design decision-making, the task of complementing the data/information is complex with the implication that a sequential process (SIRN) is required.

But here comes a difference: in pattern recognition the task is to recognize a finite pattern whereas in planning and design there is no finite pattern—it has to emerge out the planning/design process. It is here that SIRN enters the scene. Haken and Portugali have thus transformed the general SIRN model and the intrapersonal and interpersonal collective with a common reservoir submodels into decision-making models (Figs. 8 and 14).

Collective Planning

Figures 8 (*bottom*) and 14 indicate a potential—the way the third SIRN submodel can be applied to collective planning. This potential was realized by several subsequent studies that have employed SIRN as a conceptual framework and "An approach to planning discourse analysis" [30]. The latter paper was based on an empirical participatory observation conducted by Alfasi as part of her Ph.D. research [31]. In that observation she has participated in, followed and recorded, the meetings of a planning team that was preparing a plan for the city of Beer Sheva in south Israel. While the central aim of this study was to follow and expose the dynamic of the planning discourse, it also provides an empirical illustration to collective planning, namely, to the way a groups of planners are planning together.

The insight gained by this empirical study is twofold: Firstly, that a planning team can be seen as a complex, self-organizing systems the dynamics of which follows the third SIRN submodel as described above. Secondly, that discourse among the planners is the main medium through which collective planning is implemented. As further suggested recently [14, Part III], one can identify two forms of planning discourse: One that takes place between the planners who were specifically assigned to prepare the plan and another that evolves as public discourse along the lines of Habermas' *communicative action* [32] and Heley's *communicative planning* [33].

The participatory observation mentioned above exemplifies the first form, as noted. In following closely the Beer Sheva planning discourse it was possible to follow the way new planning ideas and policies emerge out of the discoursive

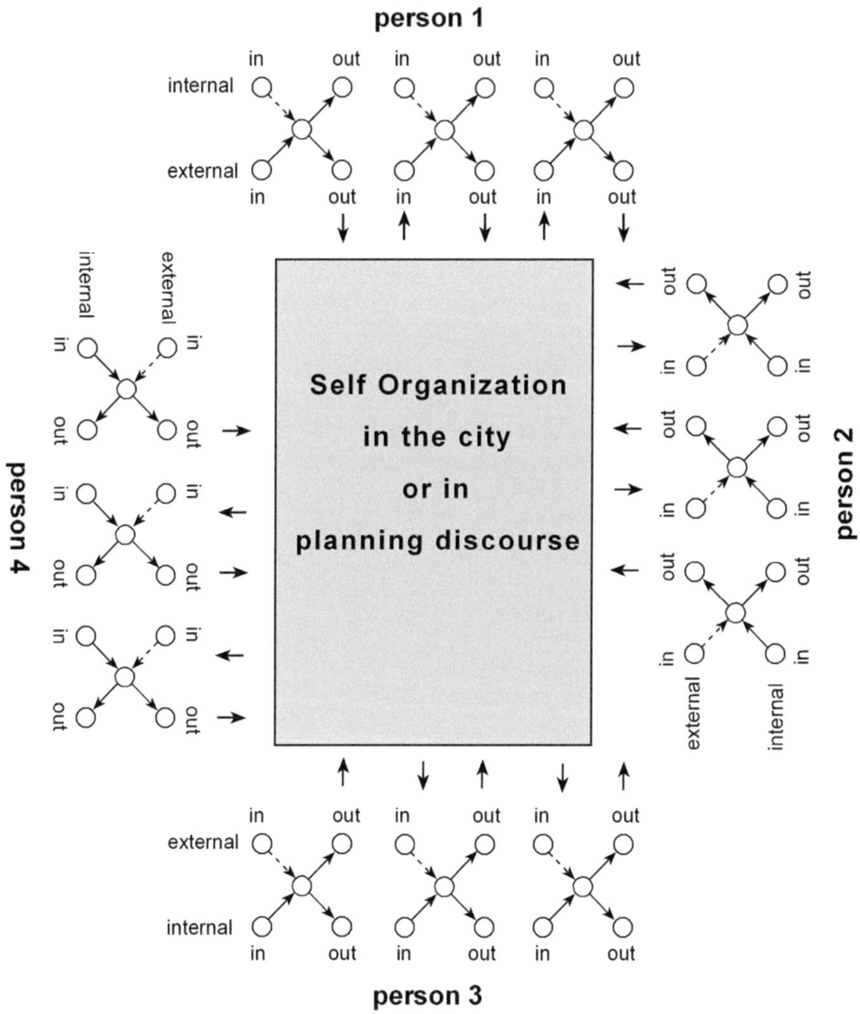

Fig. 14 Four person city planning discourse model

interaction between the various planners, how they take shape, stabilize, dominate the discourse for a certain period, just to be replaced by other ideas that emerge in the discourse and so on. This process went on until at a certain stage a given planning scheme eventually emerged as the winning order parameter that finally enslaved the discourse and brought it into a steady state during which no further plans were added to the discourse. For a detailed discussion of the actual planning discourse as it took place in the Beer Sheva team see Portugali and Alfasi [30]. In analyzing the discourse it was possible to see how various factors such as the personality and charisma of the individual planners are affecting the planning discourse and as a consequence the final result.

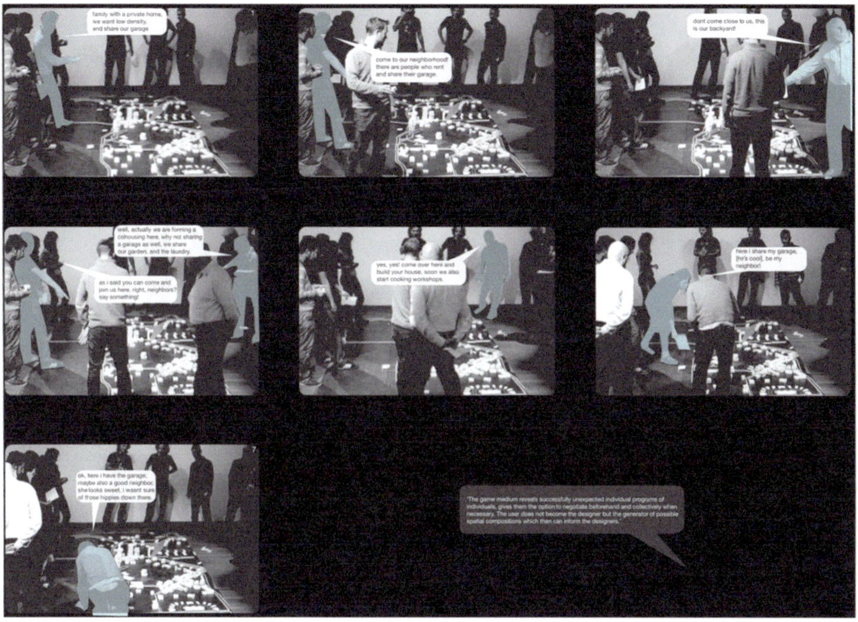

Fig. 15 Snapshots from the city planning game

Collective Design

In a recent study the *Interpersonal with a common reservoir* SIRN submodel, with its city game, were used as a framework for a design city game [34]. The game was played in the context of a real urban project: the plan to add some 350 new homes to the new town of Almere Haven, Netherlands. The Almere planning department has assigned the area of Sportpark de Wierden for the extension, and decided that the plan should be made by means of public participation. The design city game described below can thus be seen as an experiment the aim of which is to explore the usefulness of city games as a public participation design tools.

The game was thus played on a 2D map of Sportpark de Wierden when the players that simulated the new residents of Almere were fifteen graduate students with diverse cultural (Indian, American, Kenyan, Dutch, Turkish…) and disciplinary (architecture, planning, sociology, anthropology…) background. In a three-hour experiment, the participants played thirteen rounds placing mock-ups based on their resident profiles. As in previous city games, here too, the participants made location decision sequentially. However, here we've added an additional rule that 'in case of conflict, existing buildings will have priority over the new-intended ones'.

Figure 15 shows several snapshots from the game as it developed, while Figures the resultant outcome. The game was interesting in several respects. Firstly, in the sense that while it started with the two simple rules specified above, other rules came into being as emerging properties during the game; among them rules

Fig. 16 Exemplary city planning result

of development, rules of network and rules of form. Secondly, as can be seen in Figure, the resultant urban landscape is highly (self) organized and rather rich and articulated. Thirdly and in association with the above, despite the fact that there was no single mind behind the evolving urban form, the outcome is creative in the sense described below (Fig. 16).

Can the above city game be employed as an approach to urban design? My personal view is that the answer is 'Yes'! Or, more specifically, that the above game indicates a potential that has yet to be realized.

Implications to Visual Reasoning and Design Creativity

In *Knowledge of Language*: *Its nature, origin and use*, Chomsky [35] makes a distinction between external and internal languages (E- vs. I-languages respectively). E-languages are the spoken languages (Hebrew, English, French, Chinese, etc.), while I-language is the innate universal language with which, according to Chomsky, every human being comes to the world, and by means of which he or she acquires specific E-languages. Evaluating these two concepts of language from his cognitive scientific approach to the study of language he suggests that E-languages are artifacts and therefore "somewhat arbitrary, and perhaps not very interesting constructs." I agree with the first part of his suggestion, namely, that languages are artifacts; but I disagree with the second—that they are "not very interesting constructs". According to SIRN, the production of artifacts is part of cognition. Now, thinking and reasoning are implemented by means of languages—not Chomsky's I-Languages but by his E-languages—people think and reason by means of Hebrew, English, Chinese and other E-languages, that is to say, by means of artifacts.

It is true that E-languages are most dominant artifacts humans employ for thinking and reasoning, but not the only ones. People think by means of other artifacts too. From the point of view of SIRN visual reasoning is in principle not different than vocal/musical reasoning that is common in music or mimetic reasoning that is common in the domain of dance.

As we've seen above, IRN becomes SIRN when creativity enters the process of inter-representation. Namely, when the final product of the IRN process is not known in advance—not because of lack of data and information, but because it has to emerge out of the SIRN design process. From the point of view of complexity theory and SIRN creativity is therefore a cognitive phase transition—when it takes place in a single person's mind it is a result of an intra-personal SIRN process; when it is a result of collective planning it is an outcome of an inter-personal collective SIRN process.

References

1. Portugali J (1996b). Inter-representation networks and cognitive maps. In: Portugali J. (ed) The construction of cognitive maps, pp. 11–43. Kluwer Academic, London
2. Haken H (1983a) Synergetics, an introduction, 3rd Edn. Springer, Berlin
3. Haken H, Portugali J (1996). Synergetics, inter-representation networks and cognitive maps. In: Portugali J. (ed)The construction of cognitive maps, pp. 45–67. Kluwer Academic, Dordrecht
4. Das JP, Kar BC, Parrila RK (1996) Cognitive planning: the psychological basis of intelligent behavior. Sage, New Delhi
5. Lawson B (2005) How designers think, 4th edn. Elsevier, Oxford
6. Hillier B (1996). Space is the machine: a configurational theory of architecture. Cambridge University Press.
7. Alexander C (2002/2004). The nature of order: an essay on the art of building and the nature of the universe. Center for Environmental Structure, Berkeley
8. Haken H, Portugali J (2005) A synergetic interpretation of cue-dependent prospective memory. Cognit Processes 6:87–97
9. Portugali N (2006). The act of creation and the spirit of a place, a holistic-phenomenological approach to architecture. Axel Menges, Stuttgart
10. Portugali J (2002) The seven basic propositions of SIRN (Synergetic Inter-Representation Networks). Nonlinear Phenom Complex Syst 5(4):428–444
11. Varela FJ, Thompson E, Rosch E (1994) The embodied mind. MIT, Cambridge
12. Bartlett FC (1932/1961). Remembering: a study in experimental and social psychology. Cambridge University Press, Cambridge
13. Portugali J (2000). Self-organization and the city. Springer, Heidelberg
14. Portugali J (2010). Complexity, cognition and the city. Springer, Heidelberg

15. Haken H (1996) Principles of brain functioning: a synergetic approach to brain activity, behavior and cognition. Springer, Berlin
16. Miller GA (1956) The magic number seven, plus or minus two: some limits on our capacity for processing information. Psychol Rev 63(2):81–97
17. Goel V, Pirolli P (1992).The structure of design problem spaces.Cognit Sci 16:345–429
18. Gardner H (1987) The mind's new science. Basic Books, New York
19. Gibson JJ (1979) The ecological approach to visual perception, Gibson. Houghton-Mifflin, Boston, p 40
20. Haken H, Portugali J (2003) The face of the city is its information. J Environ Psychol 23:385–408
21. Shannon CE, weaver W (1949). The mathematical theory of comminication. University of Illinois Press, Illinois
22. Haken H (1988/2000). Information and self-organization: a macroscopic approach to complex systems. Springer-Verlag, Berlin
23. Attneave F (1959) Applications of information theory to psychology. Holt, Rinehart and Winston, New York
24. Krampen M (1979) Meaning in the urban environment. Pion, London
25. Arenheim R (1962/1974) Picasso's guernica University of California Press, Berkeley
26. Goldschmidt G (1991) The dialectic of sketching. Creativity Res J 4(2):123–143
27. Goldschmidt G (1994). On visual design thinking
28. Sir F. Petrie's (1904) notion of relative chronology
29. Cavalli-Sforza, Feldman (1981) have termed "cultural mutation"
30. Portugali J, Alfasi N (2008) An approach to planning discourse analysis. Urban Stud 45(2):251–272
31. Alfasi (200×) Ph.D research
32. Habermas J (1984) The theory of communicative action, vol 1: reason and the rationalization of society. Heinemann, London
33. Healey P (1997) Collaborative planning: making frameworks in fragmented societies. Macmillan, London
34. Tan E, Portugali J (2011). The responsive city design game. In Portugali J. et al (eds), Complexity theories of cities have come of age. Springer, Heidelberg (forthcoming)
35. Chomsky N (1986) Knowledge of language. Praeger, New York

Qualitative Spatial-Relation Reasoning for Design

Max J. Egenhofer

Introduction

This paper reviews *qualitative spatial-relation reasoning methods* that are applicable to design. Qualitative spatial reasoning currently extends from artificial intelligence over spatial database systems to geographic information systems. A common trait of these fields is the attempt to mimic people's, often intuitive, inferences about visual information through symbolic, logical mechanisms that do not require a graphical depiction of the spatial configuration in order to draw spatial conclusions. The inference mechanisms stress qualitative spatial properties, as they are closer to human intuition than those that use only detailed quantitative values. Three complementary aspects—formalization, conceptual neighborhoods, and compositions—form the core of spatial-relation reasoning. All three facets have been developed for topological relations between simple regions in IR^2 [1–5] as well as for intervals in IR^1 [6, 7], yielding a coherent base for spatial and temporal reasoning.

- The identification of a rationale for distinguishing different spatial relations. Contemporary qualitative spatial reasoning typically focuses on closed sets of spatial relations, which are jointly exhaustive and pairwise disjoint so that for any potential configuration there is exactly one relation in the set that describes that configuration. As such the relations become the alphabet of an abstract language to describe spatial configurations qualitatively.
- The organization of the identified relations into their conceptual neighborhoods to capture pairs of relations of highest similarity. These conceptual neighborhoods form the foundation for constraint relaxation similarity reasoning.
- The derivation of the logical inferences when relations over common objects are combined. Such inferences are particularly useful to complete incomplete descriptions or to detect inconsistencies.

M. J. Egenhofer (✉)
University of Maine, Orono, USA
e-mail: max@spatial.maine.edu

J. S. Gero (ed.), *Studying Visual and Spatial Reasoning for Design Creativity*,
DOI 10.1007/978-94-017-9297-4_9

Design analysis may benefit from qualitative spatial reasoning through its inferences to interpret design, translating from the visual to the symbolic domain, in which non-graphical explanations can be furnished to complement often-subjective arguments. Also the regularities found in the structure of spatial and temporal relations may offer new inspirations for design patterns.

The remainder of this paper first compiles some of the most relevant traits of qualitative spatial reasoning systems (Sect. 2), depicts some patterns that spatial reasoning mechanisms expose (Sect. 3), and discusses how symbolic representations of design products may be analyzed computationally with spatial-relation reasoning methods (Sect. 4).

Qualitative Spatial Reasoning Systems

Existing methods for modeling spatial relations have been comprehensively compiled in several survey articles [8, 9]. While early approaches [10, 11] addressed spatial relations in an integrated fashion, the use of tailored methods for different types—topological, direction, and metric—has prevailed during the last two decades. Current models for *topological* relations fall primarily into two major categories: (1) those based on connection [3] and (2) those based on intersection [1, 12, 13]. For two simple regions, both models yield the same topological relations (Fig. 1).

Topological Relations

The 9-intersection defines binary topological relations between two simple regions, **A** and **B**, in terms of A's interior (A°), boundary (∂A), and exterior (A^{-}) with B's interior (B°), boundary (∂B), and exterior (B^{-}) [13]. The nine intersections between these six object parts describe a topological relation and can be concisely represented by a 3×3-matrix, called the *9-intersection* (Eq. 1).

$$I_9 = \begin{pmatrix} A^{\circ} \cap B^{\circ} & A^{\circ} \cap \partial B & A^{\circ} \cap B^{-} \\ \partial A \cap B^{\circ} & \partial A \cap \partial B & \partial A \cap B^{-} \\ A^{-} \cap B^{\circ} & A^{-} \cap \partial B & A^{-} \cap B^{-} \end{pmatrix} \tag{1}$$

Topological invariants of these nine intersections (i.e., properties that are preserved under topological transformations) are used to categorize topological relations. Examples of topological invariants, applicable to the 9-intersection, are the content (i.e., emptiness or non-emptiness) of a set, the dimension, and the number of separations [14]. The content invariant is the most general criterion, because other invariants can be considered refinements of non-empty intersections. By considering the values empty ($\varnothing$) and non-empty ($\neg\varnothing$) for each of the nine intersections, one can

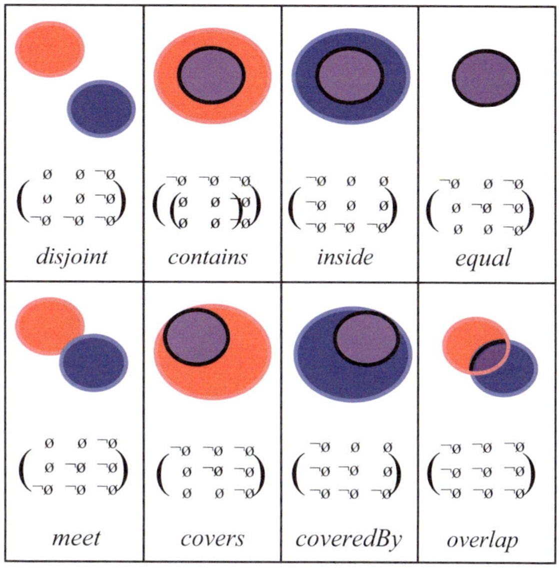

Fig. 1 The eight topological relations between two regions in $\mathbb{R}^2$ resulting from the 9-intersection and the region-connection calculus (using the 9-intersection terminology)

distinguish $2^9 = 512$ binary topological relations. Eight of these 512 relations can be realized between two regions embedded in $\mathbb{R}^2$. They are subsequently referred to as the $\mathbb{R}^2$-relations. Although the subset of the four intersections of the regions' interiors and boundaries—called the 4-intersection—is sufficient to distinguish the eight $\mathbb{R}^2$-relations [12], the 9-intersection captures critical information for making inferences about combinations of topological relations [5].

Extensions to capture more details—up to topological equivalence [14]—are relevant for relations with boundary intersections, such as *overlap* [15], *meet*, and *covers*/*coveredBy* [16]. Some alternative models for spatial relations based on interval relations [17], direction [18–20], and 2-D strings [21], have been primarily applied to image retrieval.

Often the simplicity of RCC and the 9-intesection has been cited as a particularly attractive feature, which appears to be lost when extrapolated so that all variations of more complex configurations are considered. Simplicity is a driving issue in the identification of spatial-relation models even for the most complex spatial configurations, but a formalized set of relations should not be confused with particular terminology for spatial querying. Spatial terminology of natural languages [22, 23] may be mapped onto formal models [24–27]. An effort to minimize a set of spatial relations—at the cost of suppressing, at times critical, properties—would stifle the usability of such a model; therefore, simplicity cannot always come at the cost of loosing potentially highly relevant properties.

Conceptual Neighborhoods

The enumeration of all possible relations is only the initial step, because when querying a specified spatial configuration may be unavailable in a database. In such cases the *most similar* configurations—that is, those that deviate somewhat from the exact specification [28]—should be found. For this *similarity retrieval* the relations' conceptual neighborhoods show remarkable potential.

Conceptual neighborhoods have been used successfully in the analysis of sets of relations for similarity [2, 7, 29]. The conceptual neighborhood graph captures for each relation those relations that are conceptually closest to it. Two relations are neighbors if a continuous transformation can be performed between the two relations without the need to go through a third relation. Since relations to be related typically lack a total order, their conceptual neighborhoods are used as the primary tool to provide insights about the closeness or similarity of the relations [28]. They also provide a foundation for the selection of appropriate natural-language terminology when people communicate with information systems [25].

The conceptual neighborhood for the eight topological relations in IR^2 forms a graph in which pairs of relations that are connected directly by an edge correspond to transitions that can be obtained by applying topological transformations—translation, rotation, or scaling—to one or both objects. On the other hand, pairs of relations that are not directly connected cannot be obtained through such topological transformations. The complete graph (with $4 * 7 = 28$ edges) forms the upper bound. Depending on the type of deformation that is permissible, different neighborhood graphs are obtained [30].

- Movement, rotation, and an anisotropic size-neutral deformation all yield the same neighborhoods (Fig. 2a), in which the edges that connect from *overlap* are directed.
- For an isotropic scaling all edges are directed (Fig. 2b).
- For an anisotropic scaling
- For an anisotropic scaling where the scaling applies only to some directions, all edges are non-directed (Fig. 2d).

For all types of deformations the edges from *disjoint* to *meet*, *meet* to *overlap*, *overlap* to *coveredBy* and *covers*, *coveredBy* to *inside*, and *covers* to *contains* are included in the neighborhood graphs, albeit in some cases they are directed, while in others non-directed. Transitions to *equal*, however, differ widely across the four graphs.

Inferences About Topological Relations in IR^2

The relations derived in the previous sections allow us to process topological queries in a consistent fashion, but these relations *per se* do not allow us to perform any higher-level inferences about combinations of the relations. Such combinations are of interest if a query response cannot be derived directly from the stored base relations [31]. They are also relevant to assess whether a more complex query of conjunctions of such relations can produce a result at all or whether it is internally inconsistent [32]. The latter is also useful for assessing formally whether two or

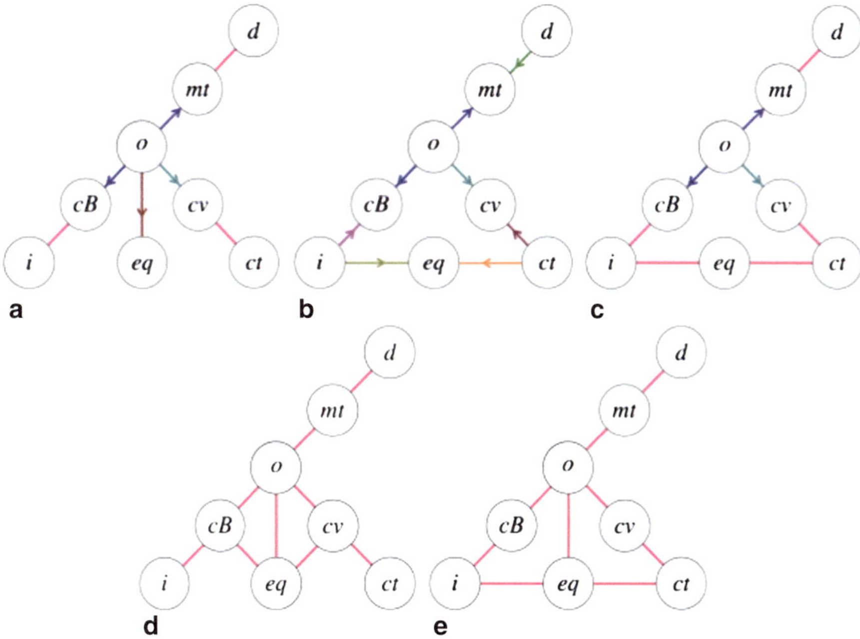

Fig. 2 The conceptual neighborhood graphs obtained from **a** movement, rotation, or anisotropic size-neutral deformation; **b** isotropic scaling; **c** anisotropic scaling; **d** anisotropic scaling with a zero-factor; and **e** the non-directed version of the union of the four graphs

more independently collected sets of spatial descriptions conform or whether they contradict each other.

Single-Relation Inferences in IR²

Some basic inferences over single relations can be made simply based on the properties of the conceptual neighborhood graph N_8 and the relations' 9-intersection matrices. Among the eight relations we find two pairs of converse relations (Eq. 2a–b), while each of the remaining four relations is symmetric (Eq. 2c–f).

$$inside(\mathrm{A,B}) \Leftrightarrow contains(\mathrm{B,A}) \tag{2a}$$

$$covers(\mathrm{A,B}) \Leftrightarrow coveredBy(\mathrm{B,A}) \tag{2b}$$

$$disjoint(\mathrm{A,B}) \Leftrightarrow disjoint(\mathrm{B,A}) \tag{2c}$$

$$meet(\mathrm{A,B}) \Leftrightarrow meet(\mathrm{B,A}) \tag{2d}$$

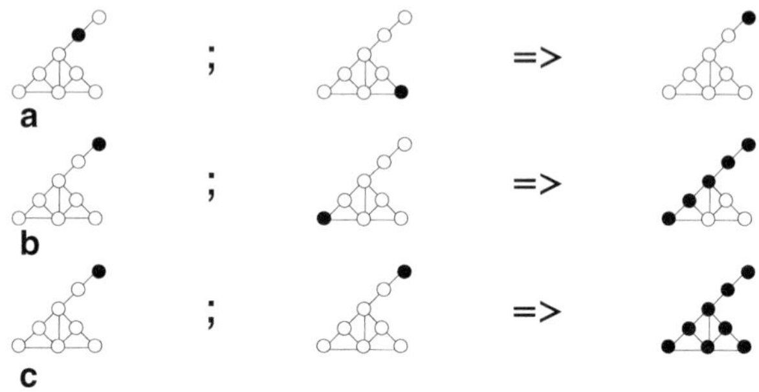

Fig. 3 Iconic presentations of compositions: **a** with a unique result, **b** with alternatives, and **c** with the universal relation as the result

$$overlap(\mathrm{A,B}) \Leftrightarrow overlap(\mathrm{B,A}) \quad (2e)$$

$$equal(\mathrm{A,B}) \Leftrightarrow equal(\mathrm{B,A}) \quad (2f)$$

Composition Table in IR²

The basis for inferences over multiple relations is the composition [33]. For a pair of spatial relations Ar_iB and Br_jC, it determines the relation (or set of relations) that may hold between A and C. Typically composition of two relations is written as r_i; r_i, omitting the references to the objects involved. For a set of n relations, the *composition table* captures all n^2 compositions.

To display the result of compositions in a compact format, we employ an iconic representation, in which each icon is based on the conceptual neighborhood graph (Fig. 2e). If a relation is part of the composition, the icon highlights it in the graph (Fig. 3a). An icon with more than one highlighted relation implies that the composition results in multiple alternatives (Fig. 3b). If all relations are highlighted, the composition of those particular relations yields the universal relation, which does not provide any inference information (Fig. 3c).

Among the 64 compositions (Fig. 4) are 27 unique compositions, four compositions with a disjunction of two relations, eight compositions with three choices, sixteen with five alternatives, and three each with six and eight choices. In non-unique compositions the relations are always connected based on the neighborhood graph, and the connections are such that they form convex shapes. These properties have a profound impact on reasoning about consistency and similarity: one may relax a relation by including its neighbors, which in turn also dilutes the possible inferences, or if one knows two or more possible outcomes of an inference, one can determine what other possibilities may occur as well.

The composition table is the foundation for assessing whether or not the spherical topological relations form a *relation algebra* [33]. Using the set-theoretic operations union (∪), intersection (∩), and complement (–), and considering *equal* as the identity relation and $\bar{r}$ as the converse relation of r, we found that all seven

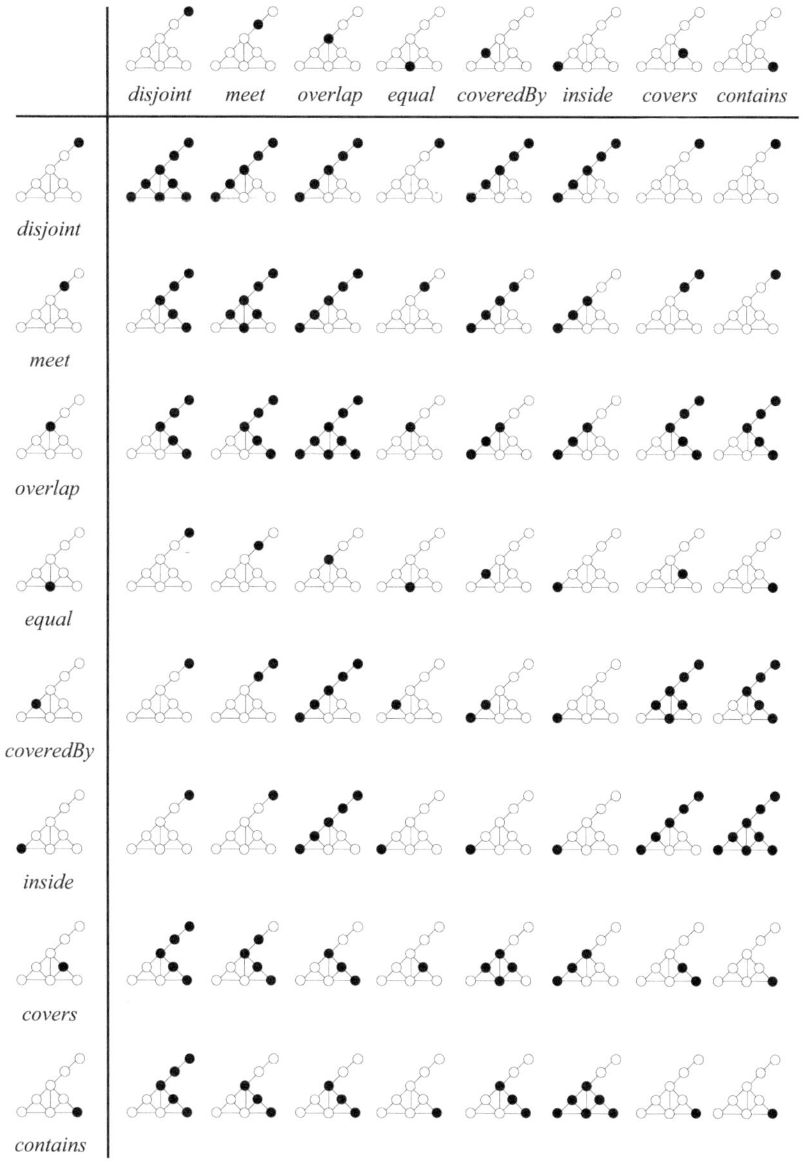

Fig. 4 The composition table of all binary combinations of the eight topological relations, displayed in their iconic presentations

properties of a relation algebra are fulfilled by the set of eleven spherical topological relations:

- Each composition with the identity relation is idempotent, because $\forall r : r;equal = r$.
- The composition with a set of relations is equal to the union of the compositions with each of the elements of the set, because $\forall (r_i, r_k) \exists r_j : (r_i \cup r_j); r_k = (r_i; r_k) \cup (r_j; r_k)$.

- The converse of a converse relation is equal to the original relation, because $\forall r : \overline{\overline{r}} = r$.
- The converse of a set of relations is equal to the union of the converse relations of each of the elements of that set, because $\forall (r_i, r_j) : \overline{(r_i \cup r_j)} = \overline{r_i} \cup \overline{r_j}$.
- The converse relation of a composition is equal to the composition of the converses of the two relations, taken in reverse order, because $\forall (r_i, r_j) : \overline{(r_i ; r_j)} = \overline{r_j} ; \overline{r_i}$.
- A variation of De Morgan's Theorem K holds, because $\forall (r_i, r_j) : \overline{r_i} ; -(r_i ; r_j) \cup -r_j = -r_j$.
- The composition is associative, because $\forall (r_i, r_k) \exists r_j : (r_i ; r_j) ; r_k = r_i ; (r_j ; r_k)$.

The major difference between the 9-intersection [13] and the region-connection calculus [4] is that for 2-dimensionally extended objects embedded in IR^2, the 9-intersection is limited to regions that are homeomorphic to 2-disks, while RCC also includes areal features with holes or with disconnected interiors. While both theories feature the same set of relations with corresponding formal definitions, the difference in the relations' domains and range (i.e., disk-like regions vs. regions with potential holes and separations) has a profound impact on the inferences that can me made reliably with the composition. For the 9-interseciton the composition is *strong*, since for each relation in a composition between two disk-like regions A and C one can find a disk-like region B such that the two composition relations hold between A and B and between B and C [34]. For some of the shapes that RCC includes, this is, however, not always possible, so that the eight RCC relations form only a *weak* relation algebra.

Coarser Topological Relations

The relation-algebra property means that one can easily form consistent disjunctions of to yield coarser relations. For instance, if for containment the difference between *inside* at the fringe (i.e., with a common boundary part) and fully inside does not matter, one can form such a relation *in* as the disjunction *inside* ∨ *coveredBy*. The consistently defined converse relation in^{-1} is then *contains* ∨ *covers*. Compositions with *in* and in^{-1} are simply the union the compositions of their constituent relations, as captured in the composition Table (Fig. 4). An often-used set of such coarser topological relations comprises the five relations that establish RCC-5, with *in*, in^{-1}, *out* = *disjoint v meet*, *overlap*, and *equal* [35], although other disjunctions have been considered as well [36]. Such relations resulting from disjunctions need not be pairwise disjoint (i.e., mutually exclusive), as for instance *equal* could be unioned with *in* as well as in^{-1}, yielding the two non-exclusive relations *inOrEqual* and $inOrEqual^{-1}$. The coarsest set of relations is *same* ← *equal* and *different* ← *disjoint* ∨ *meet* ∨ *overlap* ∨ *coveredBy* ∨ *inside* ∨ *covers* ∨ *contains*. Such a coarse set, however, would minimize the possible inferences from compositions (Fig. 5).

Fig. 5 The compositions of the two relations *same* and *different*

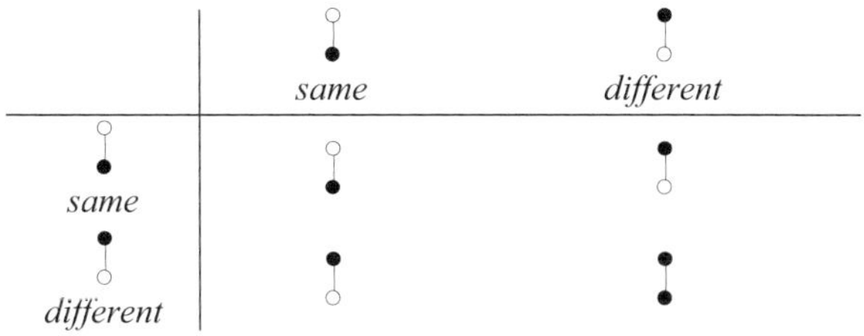

Refined Topological Relations

More details about topological relations may be captured by considering the regions' metric and directional properties. Metric refinements enhance topological relations with non-topological discernability: measures that record metric properties of common parts between two regions and measures that capture how far non-intersecting parts are from either other [37]. Together they yield nine splitting measures (Fig. 6) and two closeness measures (Fig. 7).

- Inner Area Splitting (IAS): the portion of A's interior inside of B (Fig. 6a).
- Outer Area Splitting (OAS): the portion of A's interior outside of B (Fig. 6c).
- Inverse Outer Area Splitting (OAS^{-1}): the portion of A's exterior inside of B (Fig. 6g).
- Exterior Splitting (ES): the area of A's exterior shut off by the union of A and B (Fig. 6).
- Inner Traversal Splitting (ITS): the portion of A's boundary inside of B (Fig. 6d).
- Inverse Inner Traversal Splitting (ITS^{-1}): the portion of A's interior shared with B's boundary (Fig. 6b).
- Outer Traversal Splitting (OTS): the portion of A's boundary outside of B (Fig. 6f).
- Inverse Outer Traversal Splitting (OTS^{-1}): the portion of A's exterior shared with B's boundary (Fig. 6h).
- Alongness Splitting (AS): the portion of A's boundary shared with B (Fig. 6e).
- Expansion Closeness (EC): the swelling required for A and B so that their boundaries intersect (Fig. 7a).
- Contraction Closeness (CC): the contraction required for A and B so that their boundaries intersect (Fig. 7b).

Although these metric refinements are not purely qualitative in nature, they overcome some of the constraints of purely quantitative measures, such as distances or areas captured in conventional units. The semi-qualitative nature of the metric refinements makes them more robust against scale changes, so that highly similar configurations that differ only in size would yield comparable values.

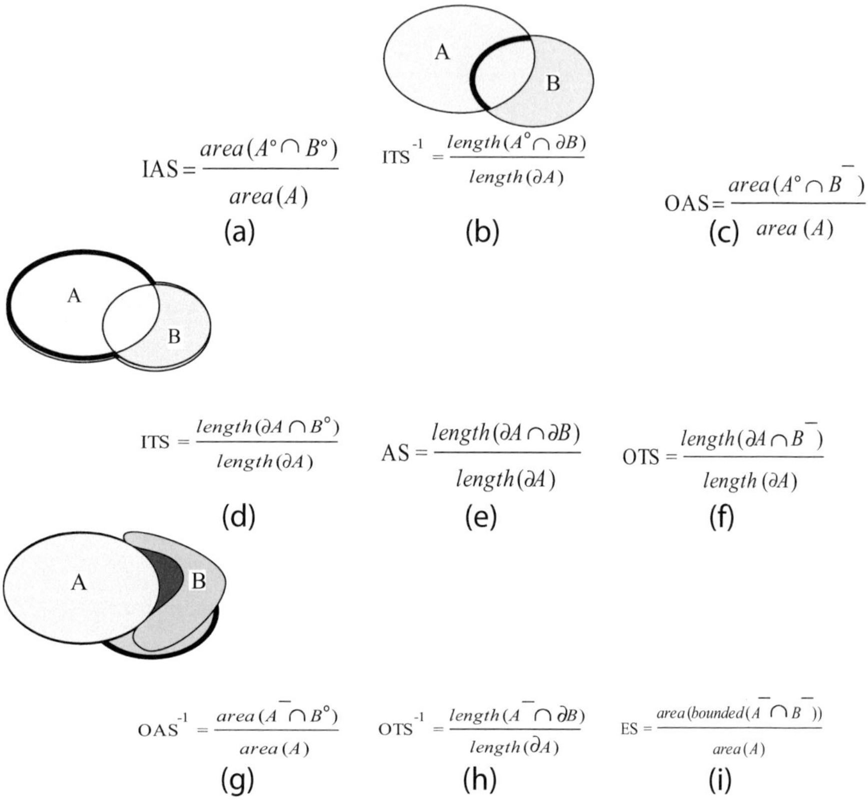

Fig. 6 The nine splitting measures [38]

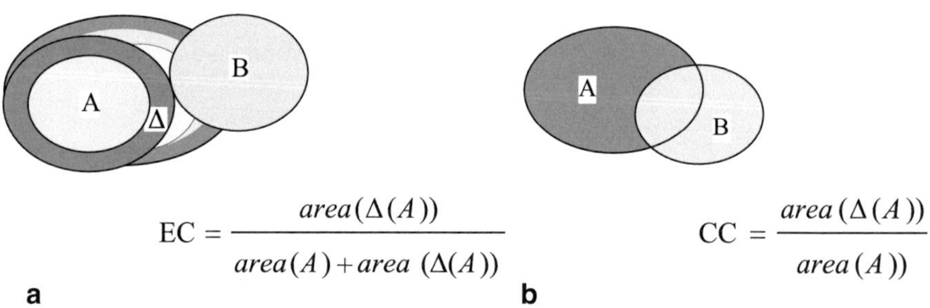

Fig. 7 The two closeness measures **a** Expansion Closeness (EC) and **b** Contraction Closeness (CC) [38]

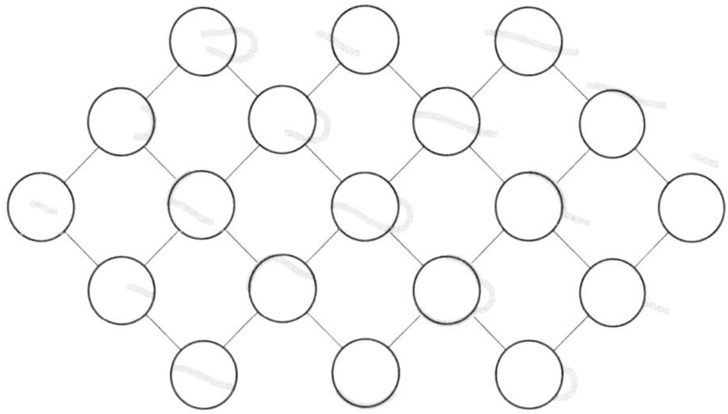

Fig. 8 The 19 line-region relations, arranged according to their conceptual neighborhoods

Topological Relations over Other Shapes than Regions

The set of realizable relations and their inferences vary depending on the types of geometric objects (i.e., their shapes) as well as the embedding space. Particularly the degrees of freedom offered by higher dimensional spaces offer a plethora of further opportunities to expand from this base set. Beyond IR^2, relations for such spaces as the real line IR^1 [6], the cyclic line SS^1 [39], the digital plane ZZ^2 [31], the surface of the sphere SS^2 [40], and the 3-dimensional space IR^3 [41] have been investigated. Here we focus, however, on other shapes in IR^2, particularly lines and holed regions.

Line-Region Relations in IR^2

The topological relations between a simple line (with exactly two endpoints and no self-intersections) and a region in IR^2 provide a richer alphabet than the eight region-region relations. The 9-intersection distinguishes nineteen different topological relations [13], whose conceptual neighborhood graph (Fig. 8) is highly regular, extending from the configuration in which the line is outside the region to the configuration in which the region contains the line in its interior [24].

These line-region relations are particularly useful for capturing the semantics of natural-language spatial predicates. It was found that to a large part the topology of a configuration is key to people's choice of a spatial preposition [25], and among the topological candidates, those that are in a particular connected subpart of the neighborhood graph are the best choices [24]. Figure 9 shows the candidate relations on the neighborhood graph that represent best the spatial predicates *crosses*, *goes into*, *goes along*, and *enters* [42].

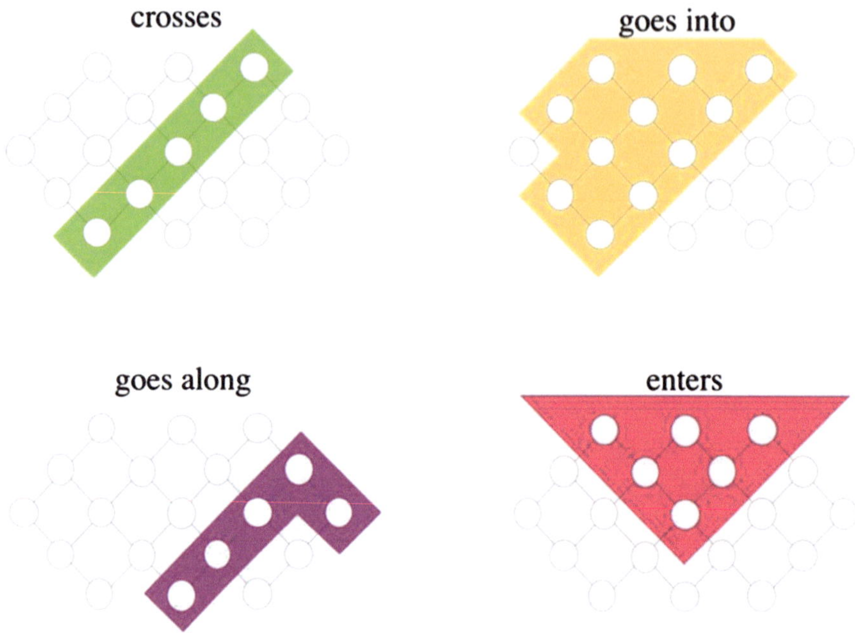

Fig. 9 The candidate relations highlighted on the line-region neighborhood for the natural-language spatial predicates **a** crosses, **b** goes into, **c** goes along, and **d** enters

Relations with Holed Regions

When a region has a hole, the number of possible qualitative relations grows from 8 to 23 [43]. These relations may be visually derived through a nested neighborhood graph, in which the first level consists of the eight relations between two regions, and the second level offers the refinements of each relation that are enabled by the hole (Fig. 10a). For instance, *disjoint* is the principal relation for one relation with a holed region, while the principal relation *inside* has eight refinements. The links among principal relations are passed on to the refined relations, yielding the conceptual neighborhood graph of the holed-region relations (Fig. 10b), which may be further transformed (Fig. 10c) to yield a more regular shape (Fig. 10d).

Qualitative Direction Relations

Directions between extended 2-dimensional objects are captured within a directional framework. In the geographic domain these direction relations are typically referred to as *cardinal directions* [45], which can be transformed into directions within an egocentric reference frame [10]. A qualitative direction relation is a triple<A, *d*, B>, where A and B are the *reference object* and *target object*, respec-

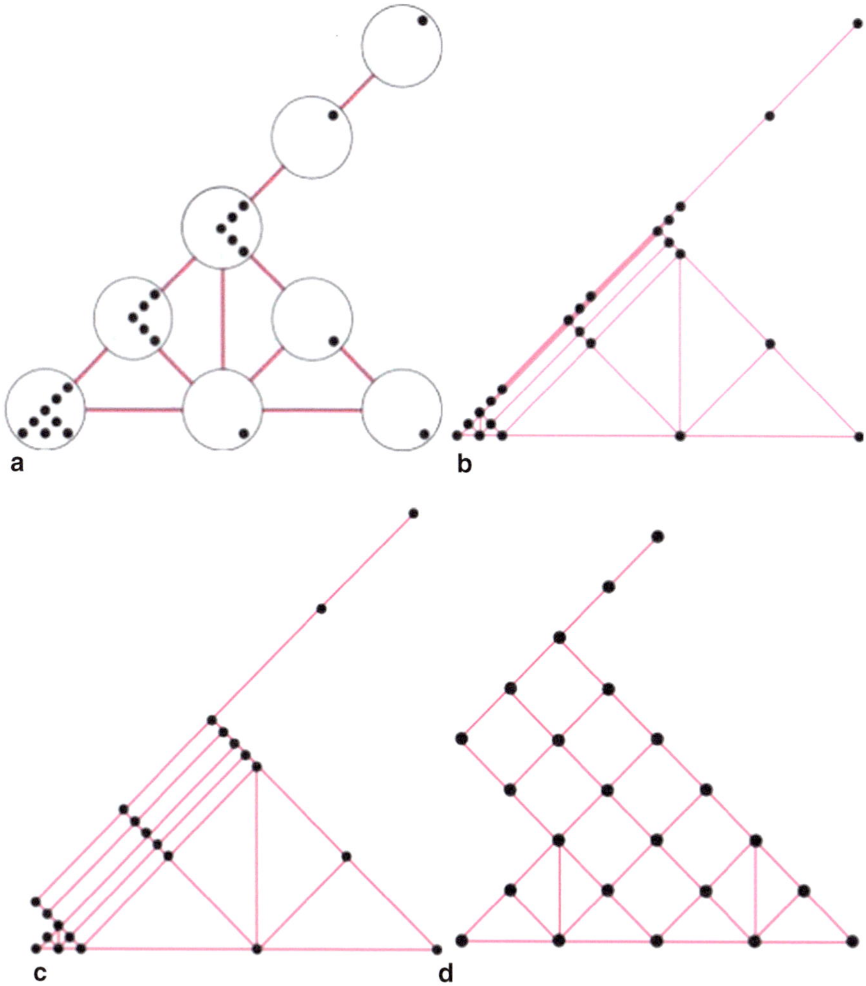

Fig. 10 The nested neighborhood graph for the relations between a region and a single-holed region: **a** nesting the refining relations inside the principal relations, **b** connections of the refining relations according to the connections of their principal relations, **c** the transformed graph so that regularities become more obvious, and **d** the rectified conceptual neighborhood graph [44]

tively, and d is a non-empty subset of the nine symbols {F, B, R, L, FR, BR, BL, BR, 0} for local directions. A direction-relation model for extended spatial objects [46, 47] considers the influence of the objects' shapes. It uses the projection-based direction partitions [45] and an extrinsic reference system [11], and considers the exact representation of the target object with respect to the reference frame. Corresponding models for internal direction relations [48] and for cone-based partitions [45] also exists [49]. The reference frame with a polygon A as reference object has nine direction tiles: front (F_A), frontRight (FR_A), right (R_A), backRight (BR_A), back

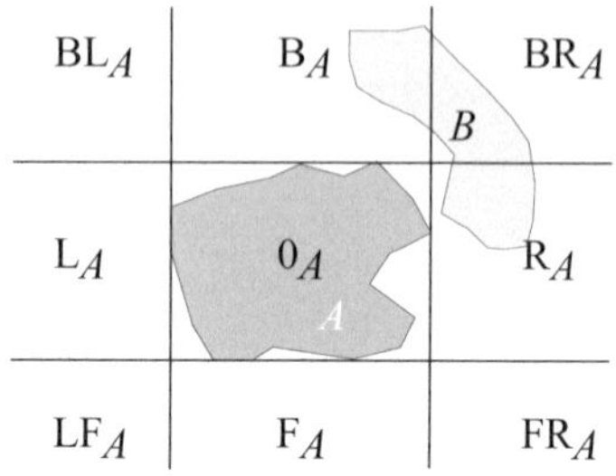

Fig. 11 Capturing the qualitative direction relation between two polygons, *A* and *B*, through the projection-based partitions around *A* as the reference object [46]

(B_A), backLeft (BL_A), left (L_A), frontLeft (NW_A), and same (0_A). The direction from the reference object to a target is described by recording those tiles into which at least one part of the target object falls (Fig. 11).

For directions between two polygons, a 3×3 matrix captures the neighborhood of the partition around the reference object and registers the intersections between the target and the tiles around the reference object (Eq. 3). The elements in the direction-relation matrix have the same topological organization as the partitions around the reference object.

$$dir(A,B) = \begin{bmatrix} \mathrm{FL}_A & \mathrm{F}_A & \mathrm{FR}_A \\ \mathrm{L}_A & 0_A & \mathrm{R}_A \\ \mathrm{BL}_A & \mathrm{B}_A & \mathrm{BR}_A \end{bmatrix} \tag{3}$$

To describe coarse direction relations, we consider the emptiness and non-emptiness of the nine intersections between the nine tiles formed around the reference object and the exact representation of the target object. Equation 4 shows the direction-relation matrix for the configuration in Fig. 11.

$$dir(A,B) = \begin{bmatrix} \varnothing & \neg\varnothing & \neg\varnothing \\ \varnothing & \varnothing & \neg\varnothing \\ \varnothing & \varnothing & \varnothing \end{bmatrix} \tag{4}$$

The matrix organization provides a rationale for the nine base relations' conceptual neighborhoods (Fig. 12).

While the empty/non-empty distinction provides for a coarse classification of qualitative direction relations, it may at times be desirable to increase the discernability beyond the empty and non-empty values. As a refinement of non-empty tile intersections the percentage of the target region's area is recorded that falls into a tile (Eq. 5).

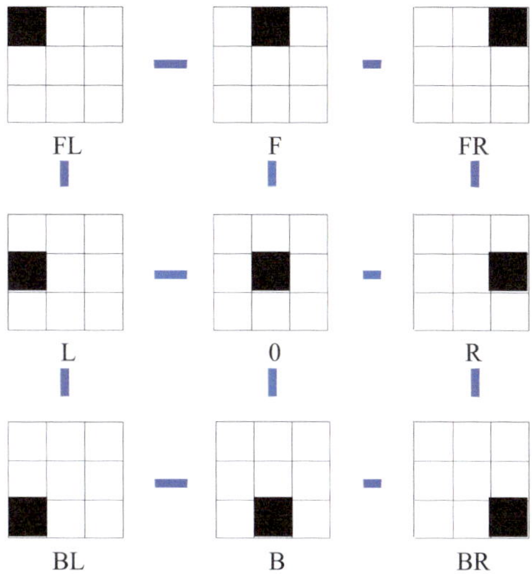

Fig. 12 Conceptual neighborhood graph of the nine direction relations [47]

$$dir(A,B)=\begin{bmatrix} \dfrac{area(\mathrm{FL}_A \cap B)}{area(B)} & \dfrac{area(\mathrm{F}_A \cap B)}{area(B)} & \dfrac{area(\mathrm{FR}_A \cap B)}{area(B)} \\ \dfrac{area(\mathrm{L}_A \cap B)}{area(B)} & \dfrac{area(0_A \cap B)}{area(B)} & \dfrac{area(\mathrm{R}_A \cap B)}{area(B)} \\ \dfrac{area(\mathrm{BL}_A \cap B)}{area(B)} & \dfrac{area(\mathrm{B}_A \cap B)}{area(B)} & \dfrac{area(\mathrm{BR}_A \cap B)}{area(B)} \end{bmatrix} \quad (5)$$

A Collection of Conceptual Neighborhood Diagrams

The geometric regularities of the conceptual neighborhood graphs have the potential to impact design by providing a collection of shapes, which have a spatial logic associated with them. The nesting of such neighborhood graphs offers then a rationale for constructing complex yet consistent diagrams [44]. Since the inner graph in such nestings may be repeated—potentially endlessly—a self-similarity construct, akin to fractals [50], may result.

While the majority of the patterns displayed so far are based on planar graphs, the planarity often does not hold for the neighborhood graphs of more complexly structured spatial objects.

The earliest conceptual neighborhood diagram [51] captured the thirteen relations between intervals in IR^1 [6]. Three different neighborhoods, depending on what deformations are considered, apply, yielding somehow different connections among the nodes. The most popular is the A-neighborhood (Fig. 13a), which fixes three of the four endpoints of the two intervals, but varies the fourth. If one cancels the intervals' orientation, implied by the underlying space IR^1, one essentially folds

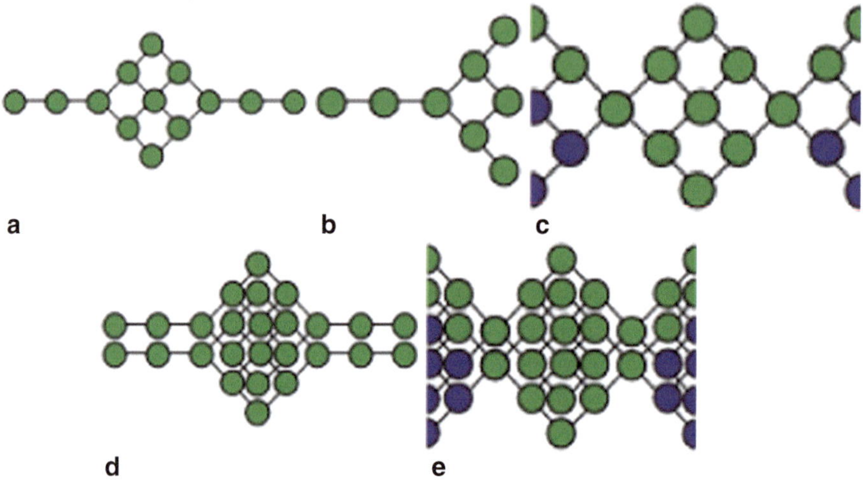

Fig. 13 Conceptual neighborhoods of interval relations: **a** intervals in IR^1, **b** intervals disregarding the orientation of IR^1, **c** cyclic intervals in SS^1, **d** oriented intervals in IR^1, and **e** oriented cyclic intervals in SS^1

the graph along its vertical axis symmetry axis, yielding a graph of eight relations (Fig. 13b) that corresponds to one of the neighborhood graphs of the eight region-region relations in IR^2. If the intervals are embedded in a cyclic one-dimensional space SS^1 one may realize 16 qualitative relations. Their neighborhood graph (Fig. 13c) is best thought of as being embedded on the surface of a cylinder, as this representation captures the repetitive nature of the cyclic phenomenon.

A further variation arises if each interval is given an explicit orientation in IR^1, which essentially doubles the number of relations. Its neighborhood graph (Fig. 13d) is then a connected pair of two A-neighborhoods for non-directed intervals. Such directed intervals can then also be projected into SS^1, yielding the corresponding doubled graph (Fig. 13e).

Similar to intervals, regions may be considered in other spaces as well. The intervals' transformation from IR^1 to SS^1 corresponds to mapping the regions from IR^2 to SS^2, yielding a highly regular neighborhood graph for the eleven spherical topological relations (Fig. 14a). Another transformation maps the regions from IR^2 into IR^3, yielding 43 region-region relations. Thirty-six of them form a two parallel graphs, each of 18 relations, which exposes a high regularity (Fig. 14b).

Such neighborhood graphs also provide a foundation for visual analyses of spatial inferences. For instance, the compositions of the cyclic interval relations in SS^1 can be categorized over the nested neighborhood graph as to the compositions' cardinality (i.e., how many base relations are part of the disjunction), with 1, 3, 5, 9, and the universal relation 12 distributing across that graph, yielding a picture of the pattern and the consistency of the distributions (Fig. 15). It shows, for instance, that all universal composition results are connected by relations of cardinality 5, and that this pattern reappears with other compositions in which 5-tuples that are connected by relations of cardinality 3.

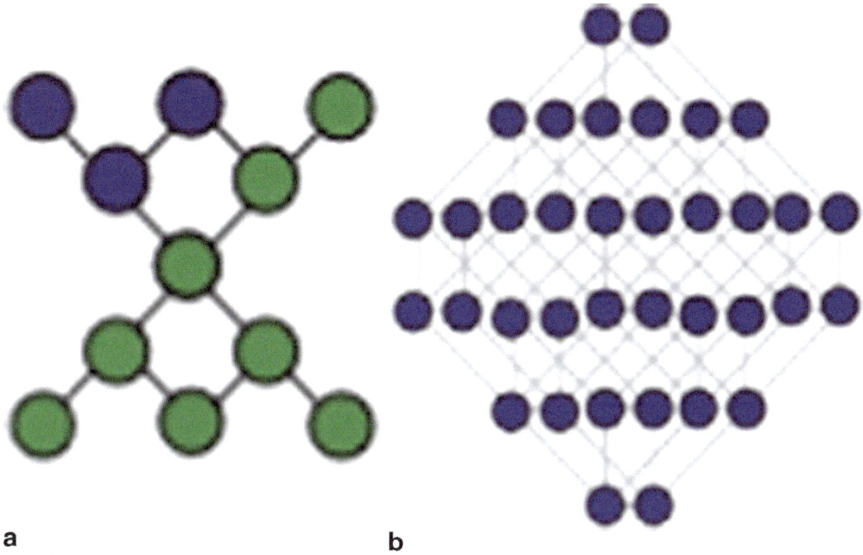

Fig. 14 Conceptual neighborhoods of region-region relations in higher dimensional spaces: **a** in SS^2 **b** in IR^3

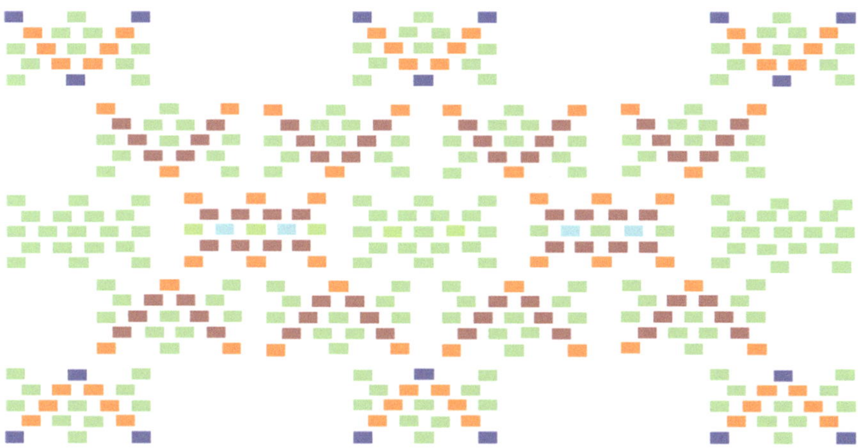

Fig. 15 Cardinality of compositions of cyclic interval relations in SS^1 visualized over their nested conceptual neighborhood graphs ([Image], [Image], [Image], [Image], and [Image])

The Role of Spatial Relations to Analyze Designs

The qualitative nature of the abstract spatial relations offers a wide range of applications. In geographic information science, the formally defined relations have become the foundation for spatial query languages. For analyzing and comparing designs, qualitative spatial relations offer additional avenues.

While graphic design is visually oriented, a complementary representation of a design product may be obtained by identifying its key features and relating them pairwise by their spatial relations. The set of binary relations—topological as well as direction—provide then the alphabet for such a symbolic descriptions of a design. Rather than analyzing visually a visual design product, the symbolic representation offers analytical, computational methods. A symbolic description of a design resembles a *spatial scene* [52], a representation that abstracts away less important aspects and focuses on the essence by capturing the key elements as well as their qualitative spatial relations. For example, *in lieu* of representing a designed graphic, with all of its details, a symbolic representation deals with such a description as "the red oval to the left of the blue line, and the yellow square inside the same red oval." The description of the key features offers an opportunity to abstract such metric details as the concrete shape of orientation of an object. Likewise, the qualitative spatial relations allow for a coarse perspective at the interplay among the objects, abstracting away primarily metric properties. For instance, the specification *inside* captures the essence of the spatial arrangement between the yellow square and the red oval, without the need to address whether the square is at the fringes or at the center of the oval, and how much of the oval's interior it covers.

Symbolic scene descriptions enable the comparison of different design products, focusing on the layout and arrangement among its elements. While visual comparisons are often highly subjective, a computational comparison of their symbolic representations resorts to the use of formally defined measures. Of particular interest in this context are similarity measures for spatial relations.

The challenging aspect of determining similarity of configurations is that most spatial relations represent discrete concepts that are usually thought of as being on a nominal scale, while similarity requires some non-arbitrary order over the elements of a scene. Similarity over different types of spatial relations is established through a metric that assesses deviations of a spatial relation from a target relation [53, 54]. The similarity assessment of an entire scene is then based on *difference measures* for each type of spatial relations. All measures are such that lower difference values represent more similar configurations, while larger values indicate more differences. A value of 0 indicates no difference according to that type of spatial relation.

Differences among topological relations are described in terms of the conceptual neighborhood graph [2]. The topological difference measure is the number of steps along the shortest path in the conceptual neighborhood graph between the target relation and the relation to be assessed for difference (Eq. 6).

$$\delta_{top}(A,B) = |top(A) - top(B)| \tag{6}$$

For example, if two objects are related by a *disjoint* relation in the target scene, while the corresponding objects *overlap* in the compared scene, then their topological difference would be 2, because it takes two steps along the conceptual neighborhood graph of topological relations to get from *disjoint* to *overlap*.

Like topological relations, directions are discrete values, which are similarly organized according to their conceptual neighborhood graph, which arranges the relations in a 3×3 grid, reflecting the nine partitions such that conceptual neighborhoods are established both in horizontal and vertical directions. The conceptual neighbors of a direction relation are then its immediate horizontal and vertical neighbors in the graph. If an object extends through more than one partition, the conceptual neighbors of its direction relations comprise the union of the neighbors of each relation in the set. The measure for coarse direction differences is for each corresponding pair of objects the shortest path between their qualitative directions within their conceptual neighborhoods (Eq. 7).

$$\delta_{dir}(A,B) = |dir(A) - dir(B)| \tag{7}$$

For example, the shortest path from B to LF is 3; from FR to BL it is 4; from B to B it is 0, from BR and W to FR and F it is 4; and from BL and L to FL it is 3.

For each type of spatial relation, a pair of scenes is evaluated for its differences in corresponding topological and directional relations. All differences are normalized so that they can be compared for difference scenes. The normalization transforms them into a value between 0 and 1, where 0 stands for "no difference" and 1 reflects the maximum difference possible.

The highest possible difference value for a pair of topological relations is 4. This value is derived from the conceptual neighborhood graph, where the longest shortest paths between any two relations is 4 (e.g., from *disjoint* to *inside*, from *inside* to *contains*, and from *disjoint* to *equal*). Therefore, the topological difference measure for a topological relation is normalized by 4 (Eq. 8).

$$\bar{\delta}_{top}(A,B) = \frac{\delta_{top}(A,B)}{4} \tag{8}$$

For an entire scene (i.e., a set of n regions), the normalized topological differences of all binary topological relations—except for those between a region and itself because they must be *equal*—are divided by the number of relations occurring (Eq. 9a). In a consistent setting, the converse relations are implied, therefore, it is sufficient to consider in the difference measure of a scene only one of the two relations for each pair. This consistency constraint reduced the amount of elements by 2 (Eq. 9b).

$$\Delta_{top} = \frac{\sum_{i=A}^{N}\sum_{j=A}^{N}\bar{\delta}_{top}(i,j)}{n^2 - n} \quad where \quad \begin{cases} i \neq j \\ n = count(A \ldots N) \end{cases} \tag{9a}$$

$$\Delta_{top} = \frac{\sum_{i=A}^{N}\sum_{j=A}^{N}\bar{\delta}_{top}(i,j)}{\left(\frac{n^2 - n}{2}\right)} \quad where \quad \begin{cases} i > j \\ n = count(A \ldots N) \end{cases} \tag{9b}$$

Differences of direction realtions are normalized by the maximum number of steps possible when moving one object onto another object along the conceptual neighborhood graph, where both objects may extend through multiple partitions. The largest possible value for a coarse direction difference is when the direction-relation matrices are populated at opposite ends. This situation requires one move at length 4; therefore, each direction difference must be normalized by 4 (Eq. 10a). Since qualitative directions do not imply their converse relations, it is necessary to consider the coarse direction differences of all relation pairs of a scene, except those relations between a region and itself (Eq. 10b).

$$\bar{\delta}_{dir}(A,B) = \frac{\delta_{dir}(A,B)}{4} \quad (10a)$$

$$\Delta_{dir} = \frac{\sum_{i=A}^{N}\sum_{j=A}^{N}\bar{\delta}_{dir}(i,j)}{n^2 - n} \quad where \quad \begin{cases} i \neq j \\ n = count(A \ldots N) \end{cases} \quad (10b)$$

The scene analysis based on topological or direction relations allows users to tailor similarity retrieval and comparisons to their needs. For example, a user whose primary interest is the preservation of topological properties may ignore in the similarity assessment the influences of directions, while a someone with a more balanced view could assign different non-zero weights to the types of spatial relations when defining similarity characteristics and perform an exploratory analysis by modifying the weights.

Conclusions

The formalisms that define sets of qualitative spatial relations and the methods to reason about them offer a rich set of tools applicable to design. The qualitative symbolic treatment of spatial relations *in lieu* of quantitative geometric representations allows for comparisons at a more abstract level than a fully detailed and complete graphical depiction. This paper highlighted some of the most commonly used sets of spatial relations and described the inference methods associated with them. Inferences include logical combinations of relations in the form of composition as well as similarity comparisons based on the relations' conceptual neighborhoods. As a side-product the shapes of the conceptual neighborhood graphs themselves yield typically highly regular figures, which may provide a motivation for design with shapes that derive from spatial reasoning.

Acknowledgments Max Egenhofer's research is partially supported by the National Science Foundation under NSF grant IIS–1016740.

References

1. Egenhofer M, Franzosa R (1991) Point-set topological spatial relations. Int J Geogr Inf Syst 5(2):161–174
2. Egenhofer M, Al-Taha K (1992) Reasoning about gradual changes of topological relationships. In: Frank A, Campari I, Formentini U (eds) Theories and methods of spatio-temporal reasoning in geographic space, lecture notes in computer science. Springer, Vol 639, pp 196–219
3. Randell D, Cui Z, Cohn A (1992) A spatial logic based on regions and connection. In: Nebel B, Rich C, Swartout W (eds) 3rd international conference on principles of knowledge representation and reasoning. Morgan Kaufmann, pp 165–176
4. Randell D, Cohn A, Cui Z (1992) Computing transitivity tables: a challenge for automated theoremprovers. 11th international conference on automated deduction, pp 786–790
5. Egenhofer M (1994) Deriving the composition of binary topological relations. J Vis Lang Comput 5(2):133–149
6. Allen J (1983) Maintaining knowledge about temporal intervals. Commun ACM 26(11);361–372
7. Freksa C (1992) Temporal reasoning based on semi-intervals. Artif Intell 54(1–2):199–227
8. Cohn A, Renz J (2008) Qualitative spatial representation and reasoning. In: van Hermelen F, Lifschitz V, Porter B (eds) Handbook of knowledge representation. Elsevier, pp 551–596
9. Galton A (2009) Spatial and temporal knowledge representation. Earth Sci Inf 2(3):169–187
10. Freeman J (1975) The modeling of spatial relations. Computer Gr Imag Process 4(2):156–171
11. Retz-Schmidt G (1988) Various views on spatial prepositions. AI Magazine 9(2):95–105
12. Egenhofer M, Herring J (1990) A mathematical framework for the definition of topological relationships. In: Brassel K, Kishimoto H (eds) Fourth international symposium on spatial data handling. pp 803–813
13. Egenhofer M, Herring J (1990) Categorizing binary topological relations between regions, lines, and points in geographic databases. Technical report. Department of Surveying Engineering University of Maine. http://www.spatial.maine.edu/~max/ 9intReport.pdf
14. Egenhofer M, Franzosa R (1995) On the equivalence of topological relations. Int J Geogr Inf Syst 9(2):133–152
15. Galton A (1998) Modes of overlap. J Vis Lang Computing 9(1):61–79
16. Egenhofer M (1993) A model for detailed binary topological relationships. Geomatica 47(3–4):261–273
17. Guesgen H (1989) Spatial reasoning based on Allen's temporal logic, technical report. International Computer Science Institute, Berkeley
18. Nabil M, Shepherd J, Ngu A (1995) 2D projection interval relationships: a symbolic representation of spatial relationships. In: Egenhofer M, Herring J (eds) Advances in spatial databases, 4th international symposium, lecture notes in computer science, vol 951. Springer, pp 292–306
19. Sistla AP, Wu C, Haddad R (1994) Reasoning about spatial relationships in picture retrieval systems. In: Bocca J, Jarke M, Zaniolo M (eds) 20th international conference on very large data bases. Morgan Kaufmann, pp 570–581
20. Sistla AP, Yu C (2000) Reasoning about qualitative spatial relationships. J Automated Reason 25(4):291–328
21. Chang SK, Shi Q, Yan C (1987) Iconic indexing by 2-D strings. IEEE Transact Pattern Anal Mach Intell 9(3):413–428
22. Talmy L (1983) How language structures space. In: Pick H, Acredolo L (eds) Spatial orientation: theory, research, and applications. Plenum Press, pp 225–282
23. Herskovits A (1986) Language and spatial cognition—an interdisciplinary study of prepositions in english. Cambridge University Press

24. Mark D, Egenhofer M (1994) Modeling spatial relations between lines and regions: combining formal mathematical models and human subjects testing. Cartogr Geogr Inf Syst 21(3):195–212
25. Mark D, Egenhofer M (1994), Calibrating the meanings of spatial predicates from natural language: line-region relations. Sixth international symposium on spatial data handling, pp 538–555
26. Ragni M, Tseden B, Knauff M (2007) Cross-cultural similarities in topological reasoning. In: Winter S, Duckham M, Kulik L, Kuipers B (eds) Spatial information theory–8th international conference, lecture notes in computer science, vol 4736. Springer, pp 32–46
27. Schwering A (2007) Evaluation of a semantic similarity measure for natural language spatial relations. In: Winter S, Duckham M, Kulik L, Kuipers B (eds) Spatial information theory–8th international conference, lecture notes in computer science, vol 4736. Springer, pp 116–132
28. Bruns HT, Egenhofer M (1996) Similarity of spatial scenes. In: Kraak MJ, Molenaar M (eds) Seventh international symposium on spatial data handling. pp 4A 31–42
29. Egenhofer M, Mark D (1995) Naive Geography. In: Frank A, Kuhn W (eds) Third European conference on spatial information theory, lecture notes in computer science, vol 998. Springer, pp 1–12
30. Egenhofer M (2010) The family of conceptual neighborhood graphs for region-region relations. In: Fabrikant S, Reichenbacher T, van Kreveld M, Schlieder C (eds) GIScience 2010, lecture notes in computer science, vol 6292. Springer, pp 42–55
31. Egenhofer M, Sharma J (1993) Topological relations between regions in R^2 and Z^2. In: Abel D, Ooi BC (eds) Third international symposium on advances in spatial databases, lecture notes in computer science, vol 692. Springer, pp 316–336
32. Egenhofer M (1994) Pre-processing queries with spatial constraints. Photogrammetric Eng Remote Sens 60(6):783–790
33. Tarski A (1941) On the calculus of relations. J Symbolic Log 6(3):73–89
34. Li S, Ying M (2002) Extensionality of the RCC8 composition table. Fundamenta Informaticae 55(3–4):363–385
35. Jonsson P, Drakengren T (1997) A complete classification of tractability in RCC-5. J Artif Intell Res 6:211–221
36. Grigni M, Papadias D, Papadimitriou C (1995) Topological inference. 14th international joint conference on artificial intelligence, pp 901–906
37. Egenhofer M, Shariff R (1998) Metric details for natural-language spatial relations. ACM Transactions Inf Syst 16(4):295–321
38. Egenhofer M, Dube M (2009) Topological relations from metric refinements. In: Wolfson O, Agrawal D, Lu CT (eds) 17th ACM SIGSPATIAL international symposium on advances in geographic information systems, pp 158–167
39. Hornsby K, Egenhofer M, Hayes P (1999) Modeling cyclic change. In: Chen PPS, Embley D, Louloumdjian J, Liddle S, Roddick J (eds) Advances in conceptual modeling: ER '99 workshops, lecture notes in computer science, vol 1727. Springer, pp 98–109
40. Egenhofer M (2005) Spherical topological relations. J Data Semant III:25–49
41. Zlatanova S (2000) On 3D topological relationships. In: 11th international workshop on database and expert systems applications, London, UK, pp 913–924
42. Shariff R, Egenhofer M, Mark D (1998) Natural-language spatial relations between linear and areal objects: the topology and metric of english-language terms. Int J Geogr Inf Syst 12(3):215–245
43. Egenhofer M, Vasardani M (2007) Spatial reasoning with a hole. In: Winter S, Duckham M, Kulik L, Kuipers B (eds) Spatial information theory–8th international conference, lecture notes in computer science, vol 4736. Springer, pp 303–320
44. Egenhofer M (2009) A reference system for topological relations between compound spatial objects. In: Heuser C, Pernul G (eds) Advances in conceptual modeling—challenging perspectives, ER 2009 workshops, lecture notes in computer science, vol 5833. Springer, pp 307–316

45. Frank A (1992) Qualitative spatial reasoning about distances and directions in geographic space. J Vis Lang Computing 3(4):343–371
46. Goyal R, Egenhofer M (2000) Consistent queries over cardinal directions across different levels of detail. In: 11th international workshop on database and expert systems applications, pp 876–880
47. Goyal R, Egenhofer M (2001) Similarity of cardinal directions. In: Jensen C, Schneider M, Seeger B, and Tstotras V (eds) 7th international symposium on advances in spatial and temporal databases, lectures notes in computer science, vol 2121. Springer, pp 36–58
48. Liu Y, Wang X, Jin X, Wu L (2005) On internal cardinal directions. In: Cohn A, Mark D (eds) Spatial information theory, international conference COSIT, lecture notes in computer science, vol 3693. Springer, pp 283–299
49. Skiadopoulos S, Sarkas N, Sellis T, Koubararkis M (2007) A family of directional relation models for extended objects. IEEE Transactions Knowl Data Eng 19(8):1116–1130
50. Mandelbrot B (1982) The fractal geometry of nature. W. H. Freeman and Co, New York
51. Nökel K (1989) Convex relations between time intervals. Proceedings 5. Österreichische Artificial-Intelligence-Tagung. Springer, Berlin, pp 298–302
52. Nedas K, Egenhofer M (2008) Spatial-scene similarity queries. Transact in GIS 12(6):661–681
53. Egenhofer M (1997) Query processing in spatial-query-by-sketch. J Vis Lang Computing 8(4):403–424
54. Li B, Fonseca F (2006) TDD—a comprehensive model for qualitative spatial similarity assessment. Sp Cognit Computation 6(1):31–62

Part III
Cognitive Science—State-of-the-Art

Thinking About Spatial Thinking: New Typology, New Assessments

Nora S. Newcombe and Thomas F. Shipley

Our world is a world that exists in space, and a world without space is literally inconceivable. Given this basic truth, it is clear that living in the world requires spatial functioning of some kind. Being creative in this world, and designing new tools and new habitats, probably requires even higher levels of spatial functioning. And people vary in their levels of spatial ability. What do these facts mean for the field of design? There are certain obvious practical questions. For example, should design schools accept only applicants who test high in spatial ability, following the lead of dental schools, which assess spatial thought on the Dental Admissions Test or with practical exercises in assignments such as tooth modeling? Or should design schools strive to enhance the spatial ability of anyone with the desire to do creative design, following the lead of selection committees for surgical residencies, which do not assess spatial ability in any way? The latter course is arguably supported by evidence (to be discussed later) showing that spatial skill is malleable. As another example of a practical question for design, consider what designers should or could know about the potential users of a product. What kinds and levels of spatial abilities should they assume that users will have? How would they be able to predict when a new tool will be too hard to master for many users, or when a building design will result in an environment in which many people easily get lost?

Sadly, this paper won't answer those very practical questions. But what it will do is provide a framework for thinking about them that differs from some prior approaches to these issues. Some writers have taken "spatial ability" to be a unitary concept—although, actually, there are literally hundreds of spatial tests, and many factor-analytic studies have suggested that there are several kinds of spatial abilities. Other writers do distinguish among various types of spatial abilities, but they do so in a wide variety of ways that do not align well with each other. Work with existing tests and statistical techniques over the past century has not arrived at a cohesive view of the structure of spatial intellect [13]. In addition (and oddly, given

N. S. Newcombe (✉) · T. F. Shipley
Temple University, Philadelphia, USA
e-mail: newcombe@temple.edu

J. S. Gero (ed.), *Studying Visual and Spatial Reasoning for Design Creativity*,
DOI 10.1007/978-94-017-9297-4_10

how many tests there are), key aspects of spatial functioning have been neglected by test makers, so that it isn't possible to define the overall structure of spatial intellect using current tools.

The field needs a new approach to this problem of defining the nature of spatial abilities. In this paper, we offer a typology for thinking about the structure of spatial intellect that derives from a top-down analysis of the nature of spatial thinking, rather than from a bottom-up inductive approach of the kind used in traditional psychometrics and factor analysis. Some of this thinking has been grounded with respect to work in the STEM disciplines (especially in geoscience), rather than with respect to design, but we think that it will ultimately inform thinking about design as well as other kinds of human endeavor. In this paper, we also point out what new assessments are needed to work with this typology.

We begin with the foundational question of what spatial thinking really is, from a theoretical and conceptual point of view motivated by a great deal of research in linguistics, cognitive science, and neuroscience. We go on to present our organizing schema, data on malleability of spatial skills derived from a recent meta-analysis, and applications of the schema to thinking about spatial skills in STEM disciplines and for identifying priority areas for developing novel assessments.

What is Spatial Thinking?

Spatial information concerns shapes, locations, paths, relations among entities and relations between entities and frames of reference. This information is represented in human cognition and can be mentally transformed to aid in manipulating, constructing, and navigating the physical world, as well as in achieving success in academic and intellectual endeavors. A variety of cognitive, neural and linguistic considerations motivate a distinction between spatial representations that are *intrinsic* to objects (their shapes and part-based representations) and those that are *extrinsic* (relations among objects and between objects and frames of reference) [e.g., 5, 8, 31]. So, for example, the spatial information that distinguishes a brush from a comb is intrinsic information, while the spatial relations between the brush and the comb and the relations of each object to the wider world are extrinsic.

Several intrinsic characteristics of an object are spatial by definition: the arrangement of constituent *parts* (sub-objects), *orientation* of the object (relative to other objects or to a reference frame), and size (or *extension)*. Transformations of these characteristics are also spatial: *bending, rotation*, and *scaling* affect arrangement of parts, orientation, and extension respectively. Another important kind of spatial transformation of intrinsic object-related spatial information involves relating 2D and 3D views (i.e., *cross-sectioning a 3D object into 2D slices* and *reconstructing a 3D structure from 2D slices*).

In addition, every object needs to be situated extrinsically—that is, it has a *location* (relative to other objects or to a reference frame). Reference frames can be either objective (or allocentric) or body-relative (and egocentric). There is an extensive literature on the processing of both allocentric and egocentric information across a wide range of species, including data on the neural substrates of the various kinds of egocentric coding and how they combine (or don't).

Useful representations of spatial information can exist at a variety of scales, and scale affects what will be defined as intrinsic and what will be defined as extrinsic. The discrete elements that we call *objects* in the natural world in which humans interact can also exist at different spatial scales, from microscopic to cosmological levels. For example, an astrophysicist may treat a galaxy as an object, whereas a microbiologist may treat a cell as an object. For any given spatial task, the appropriate scale is determined by specifying what constitutes an object (a molecule, a chair, a house, a country, a planet, a solar system, etc.). An object can, in principle, be defined at any scale, although some objects are privileged by virtue of being the entities that humans naturally manipulate in their everyday lives.

Spatial thinking necessarily involves both continuous and discrete information. Both intrinsic and extrinsic aspects of space may be coded continuously, in terms of the quantitative properties of objects and their parts (e.g., their lengths, widths) and the relations among those objects and/or parts (e.g., the distances and angles that specify relative locations). But this continuous information can also be carved up into categorical regions. Part of human understanding of space and shape involves forming categorical representations that become qualitative as they abstract away from continuous extent. For example, consider an area of low pressure on a weather map or a plateau within a mountain range. Such categorical relations are often semantically useful, as they determine whether an object is inside vs. outside another or above vs. below another. Findings in cognitive psychology on combining coordinate and categorical representations have shown the importance of spatial categories in perception and memory (e.g., [16, 17]), providing evidence that such representations are fundamental to human spatial cognition. Work in artificial intelligence has demonstrated that qualitative spatial representations are useful for modeling a wide range of engineering and scientific reasoning (e.g., [7]).

Chatterjee [5] reviews linguistic and neuroscientific data that motivates the intrinsic-extrinsic distinction that we have just introduced. He also notes that both intrinsic and extrinsic spatial relations can be represented statically but can also be dynamically transformed, either in physical actuality or through mental simulations. He discusses how these distinctions map onto linguistic distinctions, so that nouns pick out objects with distinctive shapes, prepositions refer to locations of objects relative to each other, and verbs code dynamic change, but with manner verbs referring to changes in intrinsic properties of objects and path verbs referring to changes in the locations of objects. These distinctions are realized neurally, as Chatterjee has shown in various studies of people with focal brain damage and of individuals without brain damage studied using fMRI.

The intrinsic-extrinsic distinction is supported by many other lines of research, too numerous to review completely here. One result is Kozhevnikov and Hegarty's

[22] finding of a dissociation between the intrinsic-dynamic skill of mental rotation and the extrinsic-dynamic skill of perspective taking, a distinction also supported by cognitive research dating back to Huttenlocher and Presson [17]. Further, as would be expected given our schema, the extrinsic-dynamic skill of perspective taking is more closely related to navigation skills than is mental rotation [25]. Another result is Hegarty et al.'s [15] finding of a partial dissociation between performance on intrinsic (object-based) spatial ability measures and extrinsic (environmental) measures.

There is also support for a distinction between static and dynamic skills. For example, Kozhevnikov, Hegarty and Mayer [23] and Kozhevnikov, Kosslyn and Shepard [24] found that object visualizers (who excel at intrinsic-static skills in our terminology) are quite distinct from spatial visualizers (who excel at intrinsic-dynamic skills). Artists are very likely to be object visualizers, while scientists are very likely to be spatial visualizers.

Organizing Schema for Spatial Skills

This approach to thinking about the structure of the spatial world picks out four broad categories of spatial skills. These categories give us an organizing schema for thinking about what spatial skills might be important and what skills might need to be assessed and worked on. This approach transcends prior work on ways to think about individual differences, which has been inconclusive, largely because it has been divorced from theory (as argued by Hegarty and Waller [13]). The categories are:

1. **Intrinsic-Static.** Coding the spatial features of objects, including their size and the arrangement of their parts—i.e., their configuration (e.g., to identify objects as members of categories).
2. **Intrinsic-Dynamic** Transforming the spatial codings of objects, including rotation, cross-sectioning, folding, plastic deformations (e.g., to imagine some future state of affairs).
3. **Extrinsic-Static.** Coding the spatial location of objects relative to other objects or to a reference frame (e.g., to represent configurations of objects that constitute the environment and to combine continuous and categorical information).
4. **Extrinsic-Dynamic** Transforming the inter-relations of objects as one or more of them moves, including the viewer (e.g., to maintain a stable representation of the world during navigation and to enable perspective taking).

The four cells of our typology can encompass the existing literature on spatial skills. In this paper, let's look at two very different examples of such alignment. First, the potential validity of our proposed typology is affirmed by its correspondence with an analysis by Kastens and Ishikawa [21] of the nature of spatial thinking in the geosciences. In the abstract of their paper, they write that they consider:

Table 1 Five Classes of Spatial Skills

Outcome category	Description	Examples of measures	Linn and Petersen (1985)	Carroll (1993)
Disembedding	Perceiving objects, paths, or spatial configurations amidst distracting back-ground information	Embedded figures task, flexibility of closure, mazes	Spatial visualization	Visuospatial perceptual speed
Spatial visualization	Piecing together objects into more complex configu-rations or visual-izing and mentally transforming objects, often from 2D to 3D or vice versa	Form board, block design, paper folding, mental cutting, paper folding	Spatial visualization	Spatial visualization
Mental rotation	Rotating 2D or 3D objects	Vandenberg mental rotation, cube comparison, purdue spatial visualization, card rotation	Mental rotation	Spatial rela-tions/speeded rotation
Spatial perception	Understanding abstract spatial principles, such as horizontal invariance or verticality	Water-level, water-clock, Plumb-line, cross-bar, rod and frame test	Spatial perception	Not included
Perspective taking	Visualizing an environment in its entirety from a dif-ferent position	Piaget's three moun-tains task, Guilford-Zimmerman-spatial orientation	Not included	Not included

> major tasks that professional geoscientists and geoscience learners deal with, focusing on the spatial nature of the tasks and underlying cognitive processes. The specific tasks include recognizing, describing, and classifying the shape of an object; describing the position and orientation of objects; making and using maps; envisioning processes in three dimensions; and using spatial-thinking strategies to think about nonspatial phenomena.

Kastens and Ishikawa's first task is intrinsic-static, their second task is (mostly) intrinsic-dynamic (although position may be extrinsic-static), their third task is extrinsic-static (although there may be some dynamic aspects as when a map must be rotated to align with an environment), and their fourth task is extrinsic-dynamic. Their fifth task points to an area of ability that has been widely discussed in cognitive science and neuroscience, as for example when Chatterjee [5] discusses spatial metaphor.

Second, the existing literature on the training of spatial skills can be fit into this organizing schema. In the course of a meta-analysis discussed further below, Uttal, Meadow, Hand, Lewis, Warren and Newcombe (under review) identified five classes of spatial skills on which training research has been done, and mapped them onto two prior attempts to classify spatial abilities. Table 1, taken from Uttal et al.'s

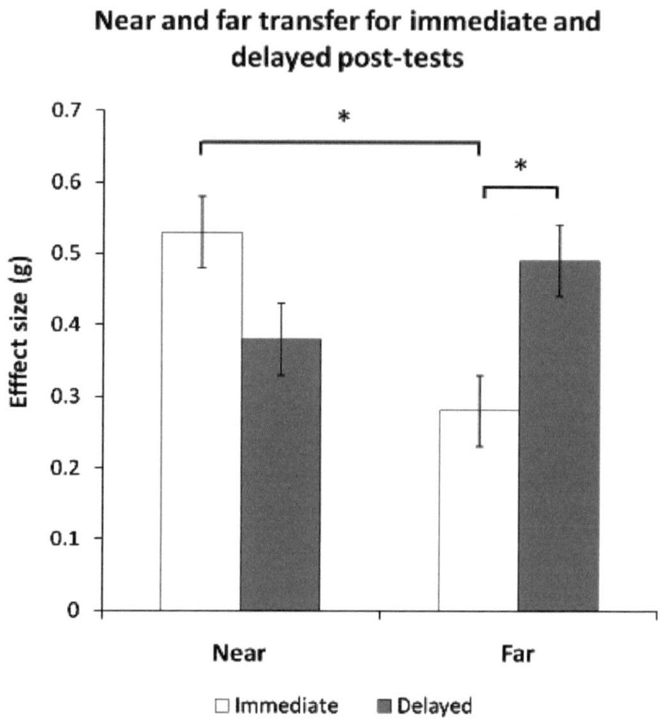

Fig. 1 Improvement after training

paper, identified an intrinsic-static skill (disembedding), two intrinsic-dynamic skills (spatial visualization and mental rotation), one extrinsic-static skill (spatial perception) and one extrinsic-dynamic skill (perspective taking). There are other examples of skills that fit into each cell, as we shall see shortly, but Uttal et al.'s aim was to encompass the existing literature. It is encouraging that a mapping was possible with studies done in the traditional psychometric tradition as well as with Kastens and Ishikawa's analysis. Such convergence among authors working in very different traditions is unusual.

Meta-Analysis to Determine Plasticity of Spatial Skills

A major meta-analysis was conducted to determine the existence and nature of plasticity in spatial skills [35]. Strikingly, all the five classes of spatial skills on which we have data show improvement after training, with effect sizes that are at least moderate in size. There is evidence that this change is lasting and that it transfers to other spatial tasks (see Fig. 1 below). Children improve somewhat more than adults, and women somewhat close the performance gap with males following training.

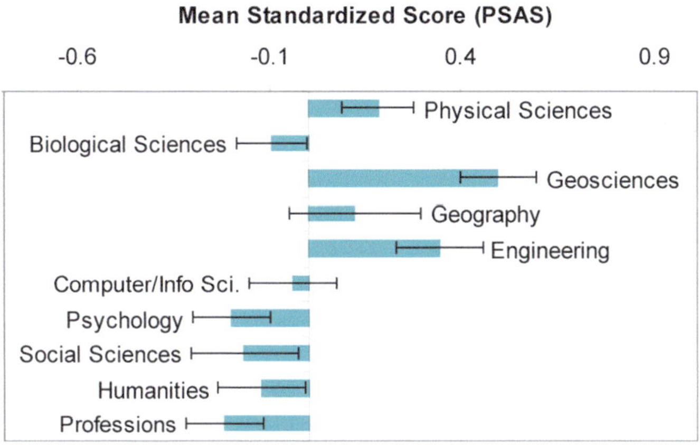

Fig. 2 Results from evaluating intrinsic skills using PSAS

Evaluating the Relevance of Intrinsic and Extrinsic Spatial Skills to Different Kinds of STEM Learning

We conducted a large self-assessment study to determine the relevance of these two broad classes of skill to STEM expertise [14]. Experts in a variety of intellectual areas rated themselves on intrinsic and extrinsic skills, using two self-report measures consisting of a set of items relating to each type of skill. Interestingly, there were different profiles of spatial skills for different STEM disciplines. In accord with prior work on predictors of STEM success (e.g., [36]), self-assessed intrinsic skills on the new Philadelphia Spatial Ability Scale (PSAS) seemed to be higher in a variety of STEM disciplines, with the exception of biology, Fig. 2.

In addition, it turned out that extrinsic skills, as self-assessed on the Santa Barbara Sense of Direction Scale (SBSOD), also showed relations to STEM expertise, Fig. 3.

This work shows that there are multiple spatial reasoning skills, and that success in some STEM disciplines requires both intrinsic spatial skills, such as those measured by mental rotation tests, and extrinsic spatial skills such as those required for navigation.

Applying the Framework of Spatial Skills to Geoscience

Successful students in geosciences require a constellation of within and between object skills for visualization of objects and for navigation. Using the framework of spatial skills discussed above, and interviews with expert geoscientists, we have identified 11 specific spatial skills used in geoscience, Table 2. Whether this set is complete, and the relative importance of these skills to STEM disciplines other than geoscience, are important open questions.

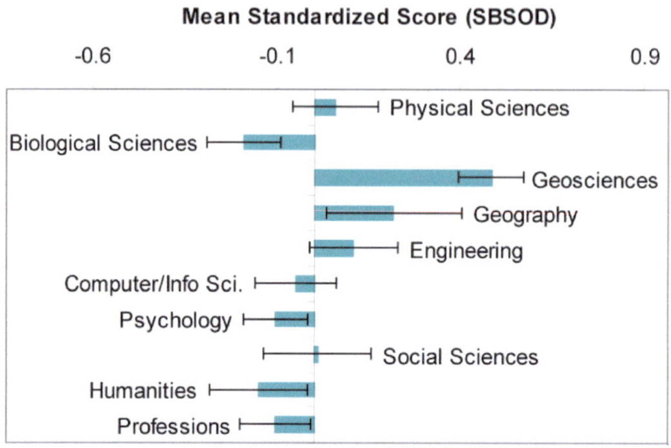

Fig. 3 Results from evaluating extrinsic skills using SBOD

Table 2 Eleven spatial skill sued in geoscience

Within-object (Intrinsic) spatial relations (2 static, 4 dynamic)
1. Disembedding: Isolating and attending to one aspect of a complex display or scene
2. Categorization: Learning categories based on spatial relations
3. Visualizing 3D from 2D: Understanding 3D spatial relations presented in a 2D image or drawing
4. Penetrative thinking: visualizing spatial relations inside an object
5. Mental transformations: visualizing how an object will change over time
6. Sequential thinking: visualizing the product of a series of transformations
Between-object (Extrinsic) spatial relations (2 static, 3 dynamic)
1. Locating self and other objects: Identifying the past or present position of objects in real space and on maps
2. Alignment: reasoning about spatial and temporal correspondence (two important cases are scaling and the use of space as a proxy for time)
3. Perspective taking: Visualizing the appearance of a scene from a different vantage point
4. Relations among objects, including self, in space: visualizing the spatial relations defined by multiple locations (e.g,. distance between 2 points and angle formed by 3 points; important for making and using maps)
5. Updating movement through space: Visualizing movement of an object relative to other objects (for self this would include route planning)

Devising Assessment Instruments for Hitherto-Neglected Geoscience Skills

To develop tools to support students that have difficulty with one or more skill, we require reliable ways to measure each student's skill level. These tests can readily be used by teachers to identify students who will need support, and by researchers to measure the effectiveness of potential interventions. However, we

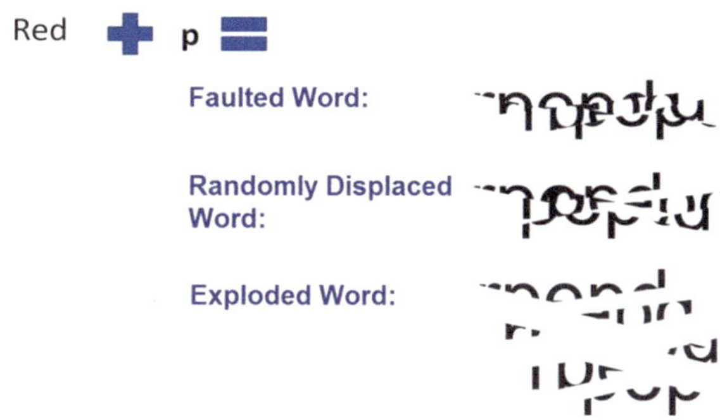

Fig. 4 Multiple test items from a sequential thinking test

have few measures for most of the spatial skills employed by geoscientists, and by extension, other STEM disciplines. We are working on test development on all spatial skills identified in geoscience practice. This will include extrinsic (or between-object relations) skills and a complete set of tests for intrinsic (or within-object relations) skills. The dual purpose of these tests leads to clear constraints: they need to be brief, easy to administer, reliable, and offer predictive and discriminative validity.

As one example of this work, Fig. 4 shows multiple test items from a test of sequential thinking [34]. Each illustrates a word (here the word is red with the letter p interleaved "r p e p d" that has been transformed by a sequence of translations of fragments of the word. In the "Faulted" case the transformations are analogous to a series of low angle faults. By reversing the multiple transformations a subject may identify the word. Interestingly, as shown in Fig. 5, this test differentiates not only between expert geoscientists and English professors, as might be expected, but also between expert geoscientists and expert chemists. Although chemists may require certain spatial skills (e.g., mental rotation), sequential thinking is not called on in chemistry and is apparently not a part of the chemist's skill set.

Devising Assessments of Individual Differences in Forming Representations of the Wider Environment

The between-object (or extrinsic) skills have been largely neglected in work on individual differences, because of the emphasis on normative questions in cognitive science and neuroscience, and on paper-and-pencil tests in psychometrics. As we have seen, a survey of STEM scientists (Hegarty et al. 2010) suggests that large-scale spatial skills may be as vital to success in some STEM disciplines as the small-scale within-object skills assessed in prior large-scale studies [36]. Navigation skills certainly seem to be critical in geosciences; for example, they allow accurate placement of field data in maps. More generally, navigation and map skills may be related and much geoscience data is presented in a structured form on a map. In

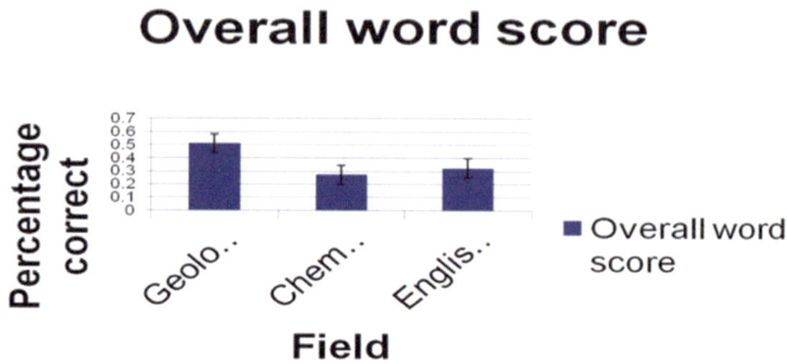

Fig. 5 Performance on multiple word tests by filed of expertise

addition, such skills seem extremely relevant to thinking about large-scale design, including architecture or landscape design. These large-scale skills may include several sub-components, such as forming integrated representations (or cognitive maps) from sequential and separated experiences, encoding the slope of the ground and incorporating slope into representations, and relating internal representations to symbolic representations, such as maps, and vice versa.

Currently, self-report measures such as the SBSOD offer basically the only practical means of assessing skills in this area, with the exception of perspective taking. We have initiated work on individual differences in representing ***extrinsic*** spatial relations. We began by confirming the results of Ishikawa and Montello (2006) showing that there are pronounced individual differences in forming cognitive maps, even in a situation in which participants walk rather than being passively driven [33]. Additionally, Schinazi et al. showed that people who formed cognitive maps showed distinctive neural activation patterns when performing a recognition tests for buildings in the learned environment. We are now building a virtual counterpart of the campus on which we ran the initial experiment, in order to provide the basis for a virtual-reality assessment tool. While Hegarty et al.'s [15] factor analysis found that learning of environments based on VR or video loaded on a different factor from learning in real environments, VR learning might still be a close-enough proxy to allow for objective assessment. Furthermore, VR learning may in some cases be closer to the learning environments in STEM disciplines than real-world learning is.

This assessment work will provide a basis for exploring how individual differences in various spatial skills affect the learning of spatial representations. For example, does better perspective taking lead to faster learning of a cognitive map? Does better mental rotation, folding, cross-sectioning, etc. lead to better scene representations (where the representations could be assessed both behaviorally and neurally)? We will also be able to examine the neural basis of categorical coding of the environment, for example, to see whether hierarchical structure revealed in behavioral studies is expressed in the hippocampus, retrosplenial cortex, or in other brain regions.

Table 3 Skills identified in early childhood

Within-Object (Intrinsic) spatial relations
1. Disembedding: Isolating and attending to one aspect of a complex display or scene. *For children, this category includes recognition of basic 2D and 3D geometric shapes. SILC is developing a test (based on Fisher et al.,* [6]*)*
2. Categorization: Learning categories based on spatial relations. *SILC is developing several tests for children of comprehension of spatial terms*
3. Visualizing 3D from 2D: Understanding 3D spatial relations presented in a 2D image or drawing. *This skill may be too difficult for young children but research has not yet evaluated whether this is true*
4. Penetrative thinking: Visualizing spatial relations inside an object. *SILC is developing a test for children* [32]
5. Mental transformations: Visualizing how an object will change when rigidly or non-rigidly transformed (e.g., moved, rotated, or folded). *SILC is developing several tests for children to supplement the Levine et al. (1999) test by adding folding (Harris) and by focusing on rotation more exclusively* [9, 32]
6. Sequential thinking: Visualizing the product of a series of transformations. *This skill may be too difficult for young children but research has not yet evaluated whether this is true*
Between-Object (Extrinsic) Spatial Relations
1. Locating self and other objects: identifying the past or present position of objects in real space and on maps. *SILC is developing several tests for children* [1, 29, 30]
2. Alignment: Reasoning about spatial and temporal correspondence (Important cases are scaling, the use of space as a proxy for time, and the use of space as a proxy for many dimensions as in graphs.) *SILC is developing several tests for children* [2, 6, 27] *(Levine et al., 2009)*
3. Perspective taking: visualizing the appearance of a scene from a different vantage point. *A test for children is needed, and SILC plans to develop one in the second five years*
4. Relations among objects, including self, in space: visualizing the spatial relations defined by multiple locations (e.g., distance between 2 points and angle formed by 3 points; important for making and using maps). *Test is needed*
5. Updating movement through space: visualizing movement of an object relative to other objects. *SILC is developing a test for children (based on Goksun et al., 2010)*

Devising Assessment Instruments Suitable for Children Under the Age of 8 Years

Early childhood is a vital period to understand, given the evidence that sex differences and SES differences appear early [26, 27] and that the utility of investment in early intervention is high [12]. Especially striking are two facts emerging from our recent meta-analysis: spatial training effects are more marked when children are younger, and yet only a very small percentage of training studies have been aimed at young children [35]. Without age-appropriate assessments, we cannot track development, or evaluate the effects of interventions. For children, we have constructed Table 3, analogous to Table 2, for skills that can be identified in early childhood.

Based on the analysis in Table 3, we have been developing assessments of skills identified as lacking in developmentally-appropriate testing instruments, including penetrative thinking [32]), paper folding [11], scaling [2, 9] and proportional reasoning [3]. These instruments are vital for allowing investigation of the sources of

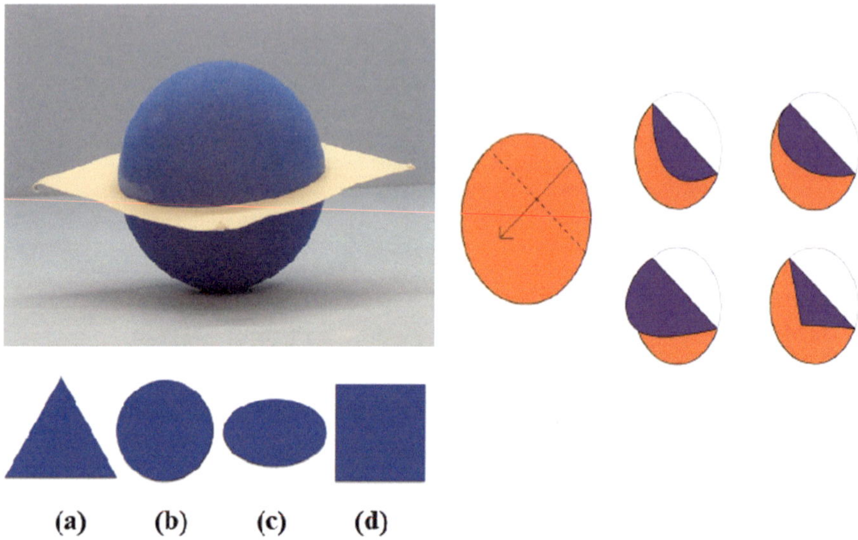

Fig. 6 Sample of spatial transformation or operation

individual differences. For example, using an available measure for young children, developed by members of our team some time ago [26], we have found that early sex differences in mental rotation relate to early use of verbal versus spatial strategies. We have also initiated work on early individual differences in representing between-object spatial relations [1] and aligning spatial patterns [10]. In Fig. 6 we give sample items from age-appropriate tests of 3-D cross-sectioning skill and of mental folding skill, respectively. In these items, children are shown a spatial transformation or operation, and are asked to select the outcome that will result.

References

1. Balcomb F, Newcombe NS, Ferrara K (2009) Convergence and divergence in representational systems: place learning and language in toddlers. In Taatgen N et al (eds) Proceedings of the 31st Annual Conference of the Cognitive Science Society. Austin, TX: Cognitive Science Society, pp 596–601
2. Boyer TW, Levine SC (under review) Child proportional scaling: is 1/3 = 2/6 = 3/9 = 4/12?
3. Boyer T, Levine SC, Huttenlocher J (2008) Development of proportional reasoning: where young children go wrong. Dev Psychol 44:1478–1490
4. Carroll J (1993) Human cognitive abilities: a survey of factoranalytic studies. Cambridge University, New York
5. Chatterjee A (2008) The neural organization of spatial thought and language. Semin Speech Lang 29:226–238
6. Fisher K, Nash B, Hirsh-Pasek K, Newcombe N, Golinkoff R (2009) Breaking the mold: altering preschoolers' concepts of geometric shapes. Poster presented at the Society for Research in Child Development Biennial Meeting, Denver, Colorado (April, 2009)

7. Forbus K, Nielsen P, Faltings B (1991) Qualitative spatial reasoning: the clock project. Artif Intell 51(1–3):417–471
8. Freksa C, Barkowsky T, Klippel A (1999) Spatial symbol systems and spatial cognition: a computer science perspective on perception-based symbol processing. Behav Brain Sci 22:616–617
9. Frick A, Newcombe NS (in prep.) Development of spatial scaling abilities. Manuscript in preparation
10. Gentner D, Lovett A, Smith L (in preparation) Tests of children's analogical ability
11. Harris J, Newcombe N (2010) Measuring spatial visualization through folding. Poster session presented at the inter-Science of Learning Center Student and Post-doc Conference, Boston
12. Heckman JJ (2006) Skill formation and the economics of investing in disadvantaged children. Science 312:1900–1902
13. Hegarty M, Waller DA (2005) Individual differences in spatial abilities. In: Shah P, Miyake A (eds) The Cambridge handbook of visuospatial thinking. Cambridge University, New York, pp 121–169
14. Hegarty M, Crookes RD, Dara-Abrams D, Shipley TF (in press) Do all science disciplines rely on spatial abilities? Preliminary evidence from self-report questionnaires. To appear in Christian Freksa and Nora S. Newcombe (eds) Spatial Cognition V. Learning, Reasoning, and Talking 5 about Space, International Conference Spatial Cognition 2010, Portland, USA. Proceedings Lecture Notes in Computer Science Springer
15. Hegarty M, Montello DR, Richardson AE, Ishikawa T, Lovelace K (2006) Spatial abilities at different scales: individual differences in aptitude-test performance and spatial-layout learning. Intelligence 34:151–176
16. Holden M, Curby K, Newcombe NS, Shipley TF (in press) A category adjustment approach to memory for spatial location in natural scenes. J Exp Psychol: Learn, Mem Cogn
17. Huttenlocher J, Presson CC (1973) Mental rotation and the perspective problem. Cogn Psychol 4:277–299
18. Huttenlocher J, Hedges LV, Vevea JL (2000) Why do categories affect stimulus judgment? J Exp Psychol Gen 129:1–22
19. Ishikawa T, Montello DR (2006) Spatial knowledge acquisition from direct experience in the environment: individual differences in the development of metric knowledge and the integration of separately learned places. Cogn Psychol 52:93–129
20. Kali Y, Orion N (1996) Spatial abilities of high-school students in the perception of geologic structures. J Res Sci Teach 33:369–391
21. Kastens KA, Ishikawa T (2006) Spatial thinking in the geosciences and cognitive sciences: a cross-disciplinary look at the intersection of the two fields. In: Manduca CA, Mogk DW (eds) Earth and mind: how geologists think and learn about the Earth. Geological Society of America, Boulder, pp 53–76
22. Kozhevnikov M, Hegarty M (2001) A dissociation between object-manipulation spatial ability and spatial orientation ability. Mem Cogn 29:745–756
23. Kozhevnikov M, Hegarty M, Mayer RE (2002) Revising the visualizer-verbalizer dimension: evidence for two types of visualizers. Cogn Instr 20:47–77
24. Kozhevnikov M, Kosslyn S, Shephard J (2005) Spatial versus object visualizers: a new characterization of visual cognitive style. Mem Cogn 33:710–726
25. Kozhevnikov M, Motes M, Rasch B, Blajenkova O (2006) Perspective-taking vs. mental rotation transformations and how they predict spatial navigation performance. Appl Cogn Psychol 20:397–417
26. Levine SC, Huttenlocher J, Taylor A, Langrock A (1999) Early sex differences in spatial skill. Dev Psychol 35:940–949
27. Levine SC, Vasilyeva M, Lourenco SF, Newcombe NS, Huttenlocher J (2005) Socio-economic status modifies the sex difference in spatial skills. Psychol Sci 16:841–845
28. Linn MC, Peterson AC (1985) Emergence and characterization of sex differences in spatial ability: a meta-analysis. Child Dev 56:1479–1498

29. Loewenstein J, Gentner D (2001) Spatial mapping in preschoolers: close comparisons facilitate far mappings. J Cogn Dev 2:189–219
30. Lourenco SF, Levine SC (2009) Location representation following early unilateral brain injury: Evidence of distinct deficits and degrees of plasticity. Paper presented at the Society for Research in Child Development Biennial Meeting, Denver, Colorado (April, 2009)
31. Palmer SE (1978) Fundamental aspects of cognitive representation. In: Rosch E, Lloyd BB (eds) Cognition and categorization. Lawrence Erlbaum, Hillsdale, pp 259–303
32. Ratliff KR, McGinnis CR, Levine SC (2010) The development and assessment of cross-sectioning ability in young children. Paper submitted to the 32nd Annual Conference of the Cognitive Science Society. Portland, OR (August, 2010)
33. Schinazi VR, Epstein RA, Nardi D, Newcombe NS, Shipley TF (2009) The acquisition of spatial knowledge in an unfamiliar campus environment. Psychonomic Society. November 2009, Boston
34. Shipley T, Manduca C, Resnick I, Schilling C (2009) Expertise in spatial visualization: can geologists reverse time? Psychonomic Society, Boston
35. Uttal DH, Meadow N, Hand L, Lewis A, Warren C, Newcombe NS (under review) Spatial training: what works, for whom and for how long?
36. Wai J, Lubinski D, Benbow CP (2009) Spatial ability for STEM domains: aligning over fifty years of cumulative psychological knowledge solidifies its importance. J Educ Psychol 101:817–835

Visual-Object Versus Visual-Spatial Representations: Insights from Studying Visualization in Artists and Scientists

Maria Kozhevnikov

This paper aims at to demonstrate that visual-object processing can support abstract visual-object representation in the same way as visual-spatial processing supports abstract visual-spatial representations, and that the visual representations contained in abstract art indeed constitute a unique and meaningful symbolic system, irreducible to that used in the visual-spatial domain. Specifically we compared how visual artists and scientists interpret abstract visual-spatial representations, such as kinematics graphs, and visual-object representations, such as modern abstract art.

Visual-Object and Visual-Spatial Imagery

Recently cognitive neuroscience has provided strong evidence that visual processing of object properties is distinct from visual processing of spatial properties. Specifically, it has been shown that higher-level visual areas of the brain are divided into two functionally and anatomically distinct pathways, the object pathway and the spatial relations pathway [1, 2]. The object (occipitotemporal or ventral) pathway processes information about the visual pictorial appearances of individual objects and scenes, in terms of their shape, color, brightness, texture, and size, while the spatial relations (occipitoparietal or dorsal) pathway processes information about the spatial relations among, and movements of, objects and their parts and complex spatial transformations. The distinction between perceptual processing of object properties versus spatial relations extends to visual mental imagery and working memory [3–5].

Despite the above behavioral and neuroscience evidence establishing the existence of visual-object processing as different from visual-spatial processing,

M. Kozhevnikov (✉)
Harvard Medical School, Charlestown, USA
e-mail: mkozhevn@nmr.mgh.harvard.edu

National University of Singapore, Singapore, Singapore

J. S. Gero (ed.), *Studying Visual and Spatial Reasoning for Design Creativity*,
DOI 10.1007/978-94-017-9297-4_11

contemporary intelligence research still retains the implicit assumption that visual-spatial processing is the only form of visual intelligence, and thus, it is the only component of visual ability included in psychometric assessments. It has been expected that this single visual-spatial ability dimension would predict performance in various professional fields that require any type of visual thinking and those individuals who have high abilities in science and math would also have high abilities in visual arts [6, 7]. However, there is growing evidence that different professional domains might rely on the use of qualitatively different types of visual processing by generating different types of visual images and manipulating them in different ways, and that visual processing of object but not spatial properties might play a crucial role in the creative processes of visual artists [8–10]. Visual artists characterize their images as typically pictorial and bright, and report preferences primarily for object imagery, while scientists characterize their images as abstract and schematic and report preferences for spatial imagery [11]. Indeed, several studies have shown that scientists surpass visual artists on visual-spatial ability tests, which required performing mental spatial transformations; while visual artists surpass scientists on tests which required generation of high resolution, pictorial images [12, 13]. Overall, these studies suggest that visual-object ability may have its own, unique predictive/ecological validity (success in visual arts and other fields that require generation of high-resolution, vivid images), irreducible to that of visual-spatial ability (e.g., success in scientific and engineering fields).

However, the prevailing view within the literature, beginning with Galton and persisting even in contemporary psychology, has been to associate visual-object processing with concrete visual thinking, low intelligence and inability to form abstract visual representations [14–16]. Pictorial visual-object processing has long been viewed as contrary, and even a hindrance, to abstract thinking in math and science domains that often require interpretation and manipulation of abstract spatial information, such as graph and diagrams [16]. Indeed, the conclusions from Galton's [17] study on imagery that "over-readiness to perceive clear mental pictures is antagonistic to the acquirements of habits of highly generalized and abstract thought" (p. 304), implied that intellectuals do not use visual-object imagery. Even in contemporary literature, following Galton's conclusions, only "intellectual" pursuits, such as mathematics and science, have been associated with the ability to form abstract representations [16]. Based on this assumption, visual art, which is rich in visual-object information, would contain only concrete representations, since the visual-object domain is unable to support abstract representations. This is doubtful, however, by virtue of the mere existence of abstract visual art. Indeed, historical analysis demonstrated that visual art might portray not only concrete visual appearances of objects and scenes, (e.g., landscapes or portraits in Renaissance art), but also represent abstract content, such as pure emotions and concepts using color and shape (e.g., Cubism and Abstract Expressionism) [9]. Thus, it still remains a question whether the representations contained in abstract visual art comprise a unique and meaningful symbolic system, irreducible to that used in the visual-spatial domain.

Thus, the major goal of our research was to demonstrate that visual-object processing can support abstract visual-object representation in the same way as visual-spatial processing supports abstract visual-spatial representations, and that the visual representations contained in abstract art indeed constitute a unique and meaningful symbolic system, irreducible to that used in the visual-spatial domain. Specifically we compared how visual artists and scientists interpret abstract visual-spatial representations, such as kinematics graphs, and visual-object representations, such as modern abstract art. Kinematics graphs provide information about motion of objects via visual-spatial schematic representations, which convey abstract concepts and relationships (e.g., position, velocity, or acceleration as a function of time). In contrast, abstract visual art provides information in a visual-object pictorial form, using colors and shapes, but at the same time conveying abstract concepts, feelings, and emotions that are not directly reflected by their literal forms. Previous research [13] has shown that scientists interpreted kinematics graphs in an abstract way, while visual artists interpreted them literally, as pictures. However, no studies had yet been conducted to compare how visual artists and scientists interpret abstract visual art information. If pictorial visual-object imagery is simply a concrete form of spatial imagery, it follows that proficiency in spatial processing, which scientists possess, would also help them interpret visual-object abstract information. If scientists are unable to do so while other individuals (artists) are, this would suggest that the visual-object domain conveys a type of abstract information utterly unique from visual-spatial abstract information.

Method

The participants were professional visual artists ($N=16$, 8 females), scientists ($N=24$, 6 females), and humanities/social science professionals ($N=23$, 14 females) who all held college degrees and had at least two years of professional experience in their fields. The group of visual artists included professional painters and designers. The group of scientists included computer scientists, physicists, biologists, engineers, biochemists, and mathematicians. Finally, the group of humanities/social science professionals consisted of historians, philosophers, linguists, English professors, and journalists.

All participants were asked to complete two tasks, the Kinematics Graph Interpretation Task and the Abstract Art Interpretation Task. *The Kinematics Graph Interpretation Task* was designed to examine differ rent approaches in interpreting spatial abstract visual information [18]. Participants viewed two kinematics graphs: Graph 1 depicted changes in an object's position over time and Graph 2 depicted changes in an object's velocity over time (Fig. 1).

The participants were then asked to visualize a real-life situation depicted by each graph, write a short description of what happened to the object that led to generation of these particular graphs, and then to draw another graph describing the same motion as the original, but translating the position versus time graph (Graph 1)

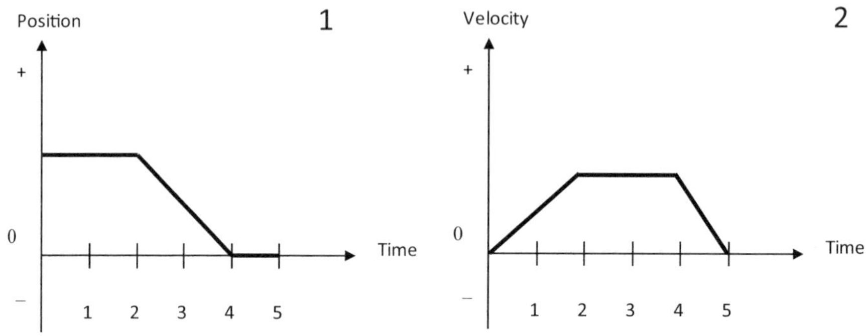

Fig. 1 Kinematics graph interpretation task: Graph 1 and Graph 2

to velocity versus time, and velocity versus time graph (Graph 2) to position versus time. The correct description of the Graph 1 is that the object is initially at rest (its position remains constant during the first time interval of the graph), then moves at a constant velocity (its position changes linearly with time during the second interval), and then at it is at rest again (its position remains constant during the final interval). The accurate drawing of velocity versus time from position versus time (translation task for Graph 1) consisted of a horizontal line at $v=0$ for the first interval, a horizontal line at a non-zero (negative) velocity for the second interval, and a horizontal line at $v=0$ for the third segment. The correct description of the Graph 2 is that the object is moving with constant acceleration (its velocity increases linearly in the first interval of the graph), then moving with constant velocity (its velocity remains the same during the second interval of the graph), and then moving with constant deceleration (its velocity decreases linearly during the last interval). The accurate drawing of position versus time graph from velocity versus time (translation task for Graph 2) had a parabolic change in position in the first step, a linear change in the second step, and a further parabolic change in the third step.

The Abstract Art interpretation task was designed to examine different approaches in the interpretation of abstract visual-object information. In this task, participants were asked to describe the meanings and feelings expressed by two abstract visual art paintings. Painting 1 (Fig. 2) is an abstract art piece by L. Berryhill, named "Breakthrough", representing the idea of liberation through adversity. Painting 2 (Fig. 2) is by W. Kandinsky, named "Kleine Welten V" (*German:* Small Worlds 5). According to art history reviews (e.g., Koehler 1998), Kandinsky represented in this painting a plan for a utopian city, and the life within, in both a physical and metaphorical sense, i.e., the "small world" with its soul and dynamics. In Painting 1, emotions expressed in the piece are much more violent and easily accessible than in Painting 2, due to its complementary splashes of color which extend beyond the scope of the picture frame.

The participants were presented with these paintings without the names of either the artists or the works. All the participants indicated that they had never seen these paintings before. The participants answered the following questions: "*What does*

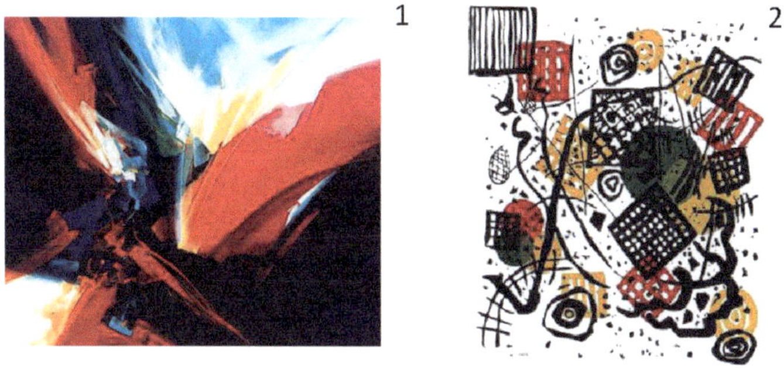

Fig. 2 Abstract art interpretation task: Painting 1 and Painting 2

this picture bring to your mind? What is drawn here? What mood/emotions/feeling does this picture evoke? How would you title this picture?" In addition, participants were asked to judge these two paintings on a 7-step bipolar scale of *emotional versus rational*, as well as a 5-step bipolar scale of *dislike versus like*.

Results: Graph Interpretation Task

For the Graph Interpretation Task, consistently with the previous categorization [18], participants' responses were coded into three categories according to the degree to which they reflected the abstract content of the graph: (a) *literal pictorial interpretations*, in which participants interpreted the graph literally, as a pictorial illustration of a situation and when translating a graph into different ordinate, they replicated the literal shape of the original graph, (b) *irrelevant interpretations*, in which participants interpreted the graph in terms of irrelevant, but not pictorial, features of the graph, and their translation reflected little to none of the information given in the original graph, or (c) *abstract schematic interpretations*, in which participants referred to the graph as a non-literal spatial representation of movement over time (independent of whether the actual interpretation was correct or incorrect) and when translating a graph into a different ordinate axis they attempted to reflect the stepwise changes in motion meaningfully. Most participants' interpretations of the graphs were highly consistent with their translations of a graph into a different ordinate axis in the sense that the responses to both components of the task fell into the same category. Only very few participants (6.56 %) produced translations of one graph to another that were inconsistent with their description of the original graph (in these cases, responses were coded into the category that less reflected the abstract content of the graph).

Two independent raters analyzed the participants' responses to the graph interpretation task, with an inter-rater reliability of 0.93. The examples of typical abstract/schematic, pictorial and irrelevant responses are given in Fig. 3.

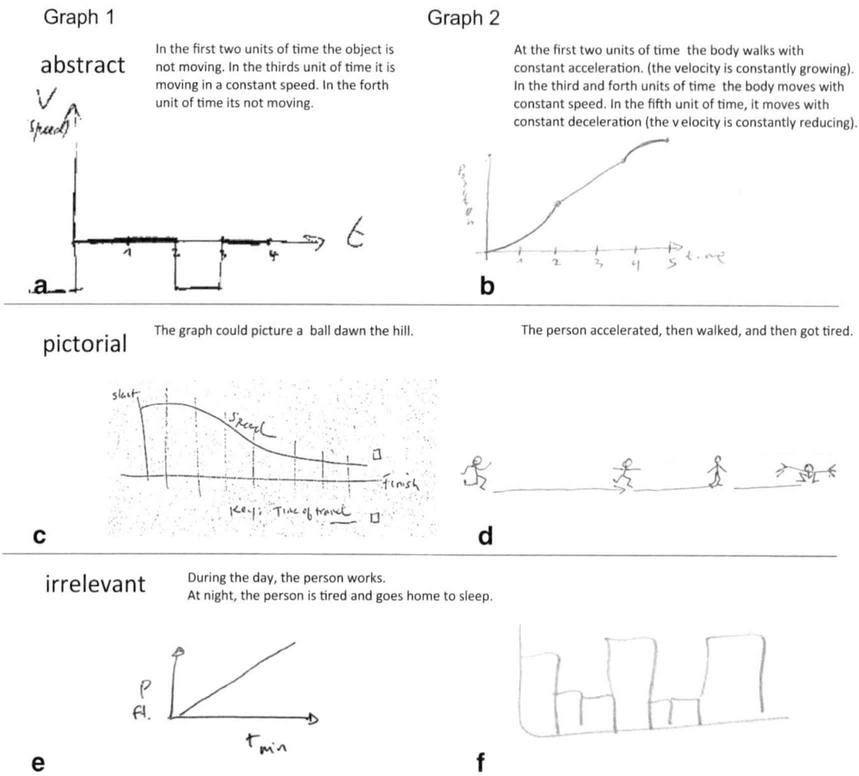

Fig. 3 Examples of different types of interpretations of the graphs

Most of the *scientists* (Graph 1: 70.8 %; Graph 2: 75.0 %) produced abstract/schematic interpretations of the object's movement represented on both graphs, while most of the *visual artists* (Graph 1: 93.8 %; Graph 2: 56.2 %) produced literal/pictorial interpretations of the object's movement. Most of the scientists considered the visual information provided by the graphs part-by-part, and described the object's motion in steps (i.e., "*1. object at rest; 2. it suddenly starts motion with constant velocity; 3. it suddenly stops and no motion again* for Graph 1, or—*there is a permanent motion of an object having 3 steps:(1) speeding up; (2) motion at constant speed, (3) slowing down; position of object is changing constantly until its full stop* for Graph 2").

In contrast, most visual artists interpreted both graph problems as literal pictorial illustrations of a situation or the object's trajectory, and did not attempt to interpret the graphs as abstract schematic representations. The artists consistently referred to the global shape of the graph and expected the shape of the graph to resemble the path of the actual motion. For both graphs, the pictorial literal interpretations given by the majority of artists emphasized the shape of the graph (e.g., "*a plane landing at airport, a squirrel climbing down a tree*" referring to the downward slope of

Graph 1, "*the balloon was flying up, moving forward, and then going down*" for Graph 2), and in some cases, the shape of the graph was interpreted literally, but in the horizontal plane (e.g., "*car is changing lanes*"). Consistently with their verbal interpretations, their drawn interpretations converting position versus time to velocity versus time for Graph 1 usually replicated this descending trend, or provided a pictorial illustration of the verbal descriptions (e.g., a drawing of ocean waves where the boat was lowered, or a drawing of an airplane and its descending trajectory). Similarly, for Graph 2, visual artists provided mostly literal interpretations, and created drawings that replicated the shape of the original graph.

Humanities/social science professionals provided a comparable proportion of answers in each category. They had the greatest number of irrelevant answers, which consisted of vague and general descriptions unrelated to the graphs (e.g., *the weather is quiet and calm, the wind is 3–5 m/s, the wind is favorable*), as well as metaphorical descriptions ("*someone died, the end of the wash cycle*").

Chi-square analysis revealed significant differences between professionals' interpretations of Graph 1 ($chi^2(4)=35.439$, $p=.001$) and Graph 2 ($chi^2(4)=28.459$, $p=.001$) (see Figs. 4a and 4b).

Results: Abstract Art Interpretation Task

A coding system consistent with that used in the graph interpretation task was applied to the art interpretation task. Responses to questions about the meaning and emotional content conveyed by the paintings were classified into three principal qualitative categories: (a) *literal/pictorial interpretations*, in which participants interpreted the painting in terms of its surface features, such as colors or concrete objects resembling the shapes in the paintings, and indicated superficial or lack of emotions (b) *irrelevant interpretations,* in which participants descriptions were irrelevant to the painting's appearance or emotional content, vague and confused, or missing entirely (c) *abstract/conceptual interpretations,* in which participants referred to the paintings in terms of conceptual and emotional content that was not directly depicted but was related to the ideas expressed by artists.

Responses were rated by two coders, and inter-rater reliability was 0.94. All participants indicated that they had never seen these paintings before. Chi-square analysis revealed significant differences between professionals' interpretations of Painting 1 ($chi^2(4)=21.714$, $p<.001$) and Painting 2 ($chi^2(4)=23.443$, $p<.001$) (see Figs. 5a and 5b).

Contrary to the widespread notion that pictorial information does not support abstract representations, most visual artists provided abstract, rather than literal-pictorial interpretations of visual information. For Painting 1, visual artists (62.5 %) provided interpretations that contained the ideas of breakthrough in both conceptual and emotional meaning (*crash and liberation, breakthrough, eruption, war, catastrophe, extreme tension*). Most of the visual artists included emotional attributions in their descriptions of art, even when simply describing the content or naming the

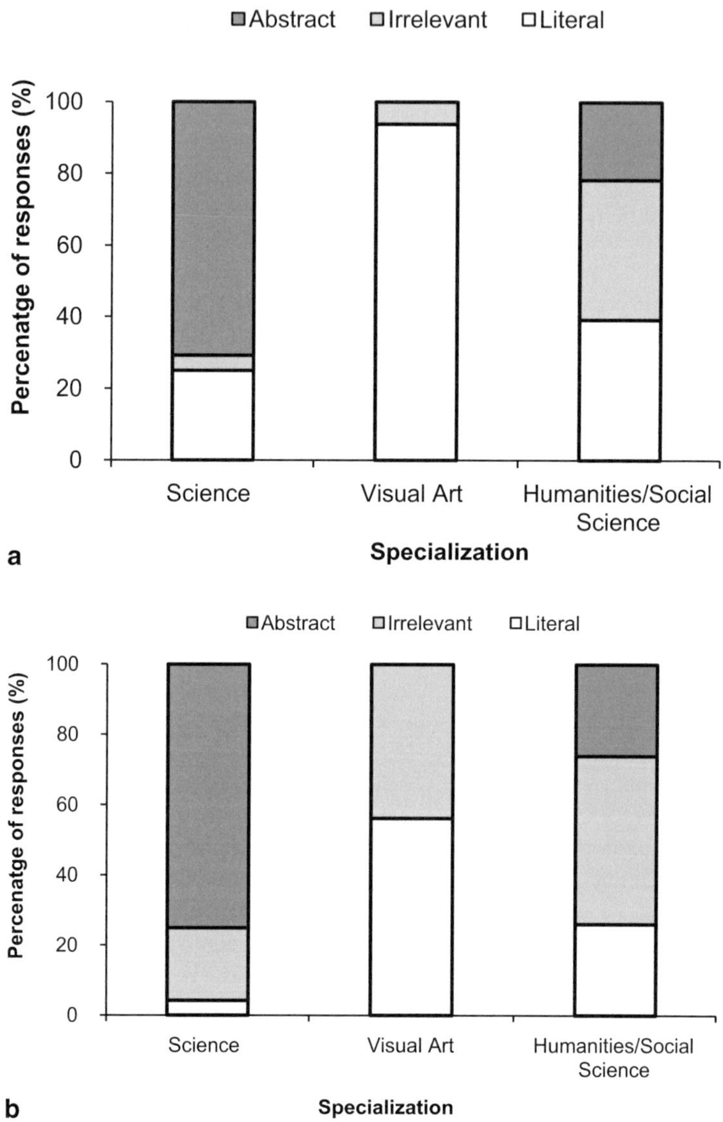

Fig. 4 Abstract interpretations of the kinematics graphs by members of different professions: **a** Graph 1 and **b** Graph 2

picture. They expressed intense and complex emotions, which in fact correspond to the concept of "breakthrough" (e.g., *explosion of emotions, fear, anxiety, horror, bursting, alarm, disturbance, extreme tension*).

Similarly, for Painting 2, most visual artists (75.0 %) provided interpretations that attributed meanings reflecting dynamics, complex life, an especially city life and city landscape (*a city, sort of abstract version of movement through a city lights, buildings, a joyful representation of a craziness of city night, a city, a loud party,*

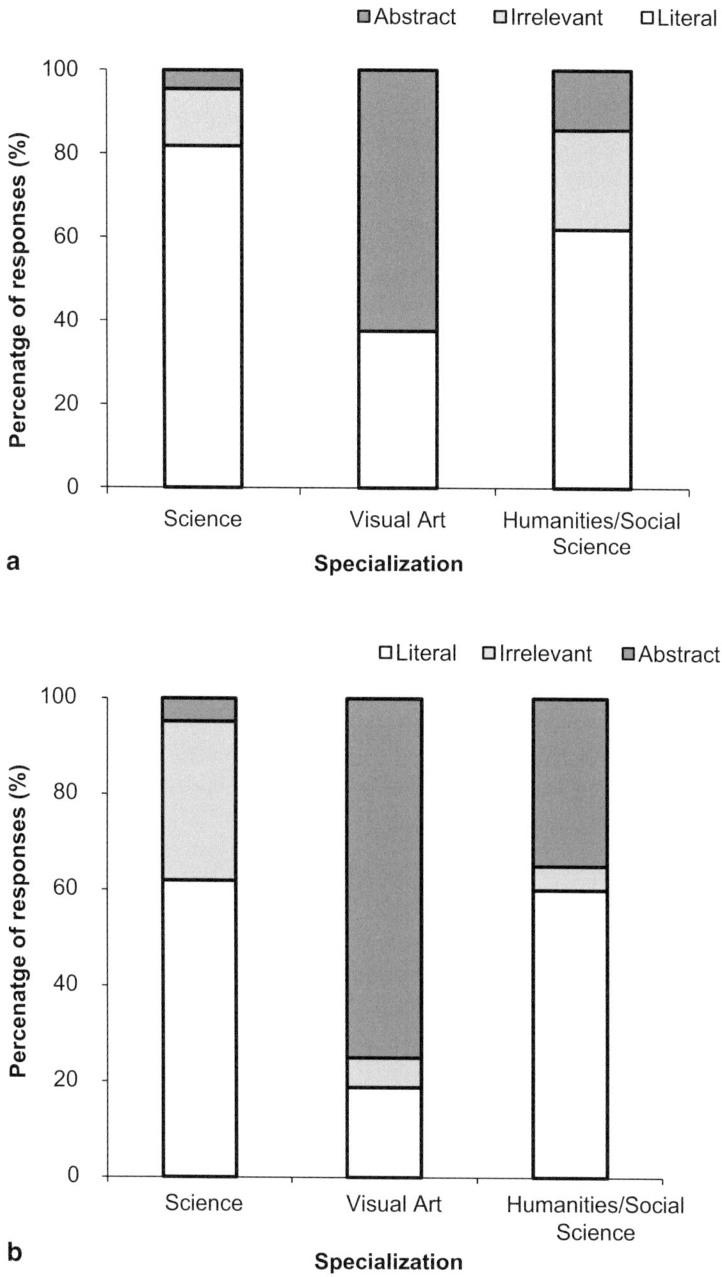

Fig. 5 Interpretations of abstract art by members of different professions: **a** Painting1 and **b** Painting 2

a whole bunch of different things but working together, moving together, party, hip hop in life, streets, tracks, theater, the Earth and people's thoughts above it).

In contrast to visual artists, 81.8 % of scientists provided literal/pictorial interpretations to Painting 1 and 61.9 % of scientists provided literal/pictorial interpretations for Painting 2. Their responses reflected surface characteristics of apparent visual features (e.g., *different colors: blue, black, red, yellow, white; sharp edges in red* for Painting 1; *jumble of color* for Painting 2. Moreover, they tended to give descriptions demonstrating understanding of content as random, non-meaningful combinations of lines, shapes and colors (e.g., *mess, trash, some shapes, no order, stains, squares, cubes, circles* for Painting 1; *conglomeration of swatches* for Painting 2 or attempts to synthesize the patterns in the paintings into concrete objects, ("*crystals of ice; pieces of ice, glass, mountains* for Painting 1; *confetti at the party, blanket design with some random pattern*" for Painting 2. Typically, in Painting 1, scientists recognized crystals or pieces of colored glass, which were not usually accompanied by any emotional descriptions, or they recognized mountains, which were usually associated with a positive and relaxed state and mood, as if they were walking on a nature trail. Scientists expressed fewer feelings overall than artists; half of the scientists did not use any emotional expressions: they either expressed no feelings at all (e.g., *nothing, no feelings),* gave estimates/descriptions of the picture itself (e.g., *beautiful scenery, nature)*, or made responses irrelevant to emotional descriptions (e.g., *everything happens together* for Painting 1; *there is a lot to be done here, this will take a lot of work, thinking about something, don't like, mess* for Painting 2.

Only a relatively small percentage of humanities/social science professionals gave abstract interpretations to paintings (14.3 % for Painting1 and 35.0 % for Painting 2, The majority of humanities/social science professional tended to provide either irrelevant answers or literal interpretations (*don't see anything, this is an abstract painting, complicated, don't want to figure out, abstract, nothing, bad drawing, It brings to mind the idea that people may in fact create art for no reason. I think if there is a message or emotion behind this creation that's too esoteric for anyone the outside world to understand*). Interestingly, humanities/social science professionals tended to come up with broad digressive metaphors, instead of emotional descriptions (e.g., *excitement & diversity, a clash of order & chaos, ecstasy of mind*).

Discussion

Overall, the results from both the *Abstract Art interpretation task* and *Graph interpretation task* showed that, visual artists and scientists tended to interpret abstract visual information in qualitatively different ways from each other, depending on the domain of visual information. Visual artists tended to interpret abstract art as abstract representations, but scientists and humanities/social science professionals tended to interpret abstract art literally, in a concrete way. In contrast, visual artists tended to interpret graphs literally (graphs-as-pictures), but scientists tended to interpret graphs schematically, in abstract way, while humanities/social science professionals did not have any clear tendency to either type of interpretation. Compared to other professionals, humanities/social science professionals gave the most irrelevant responses, possibly because they may have been confused by the tasks, or that they attempted to solve them by purely verbal-analytical approaches.

The results demonstrated that visual artists were indeed able to extract abstract information from abstract representations in the visual-object domain, and that their interpretations were often consistent with the interpretations by other artists and the intent of the painters, while scientists and humanities/social science professionals were not. This indicates that visual artists have the capacity to use an abstract symbolic system unique to visual-object domain, and to infer meaningful information from visual-object abstract representations. The fact that the artists have experience in dealing with such representations is immaterial; if they have learned to interpret art in such a way, this indicates that they have learned the abstract "language" of the visual-object domain, in the same way that scientists have done so in the visual-spatial domain, and by necessity, this proves that abstract processing in the visual-object domain does indeed exist. Furthermore, since the results indicate that proficiency in visual-spatial abstract processing (either due to experience or inborn capabilities) does not help scientists interpret visual-object information abstractly, proficiency in visual-spatial processing is not sufficient for supporting abstract representations in the visual-object domain. Thus, we conclude that visual-object imagery cannot be considered a concrete form of visual-spatial reasoning, but constitutes an independent domain that supports its own abstract visual-object representations.

Furthermore, the results show that humanities/social science professionals do not necessarily form abstract representations in either the visual-spatial or visual-object domain, contrary to the view that characterizes those strong in abstract verbal processing as more inclined to think abstractly in general [19, 20]. Therefore the current results suggest that abstract verbal thinking may not be beneficial for visual-object and visual-spatial tasks, and that abstract thinking is domain specific (i.e., verbal, object or spatial). Another important inference from the current study is that visual-object, but not visual-spatial, processing may be related to emotional attribution; visual artists expressed a richer, deeper range of emotional reactions, relevant to the content of the paintings, compared to the other two groups.

Overall, these results, contrary to the claims of Galton and his followers [16], suggest that pictorial visual-object imagery does not impede abstract thinking, but in fact supports a different type of abstract thinking. Furthermore, the results of our study demonstrate that the representations contained in abstract art indeed comprise a unique symbolic system, whose content is consistently meaningful and accessible across individuals, and is irreducible to that used in the visual-spatial domain.

References

1. Ungerleider LG, Mishkin M (1982) Two cortical visual systems. In: Ingle DJ, Goodale MA, Mansfield RJW (Eds), Analysis of visual behavior. MIT Press, Cambridge
2. Kosslyn SM, Koenig O (1992) Wet mind: the new cognitive neuroscience. Free, US
3. Farah MJ, Hammond KM, Levine DN, Calvanio R (1988) Visual and spatial mental imagery: dissociable systems of representations. Cognit Psychol 20:439–462

4. Kosslyn SM (1994) Image and brain: the resolution of the imagery debate. MIT Press, Cambridge
5. Mazard A, Tzourio-Mazoyer N, Crivello F, Mazoyer B, Mellet E (2004) A PET meta-analysis of object and spatial mental imagery. Eur J of Cognit Psychol 16:673–695
6. Eisner EW (1985) The art of educational evaluation: a personal view. Falmer, London
7. Gardner H (1999) Intelligence reframed: multiple intelligences for the 21st century. Basic Books, New York
8. Kassels S (1991) Transforming imagery into art: a study of the life and work of Georgia O'Keeffe. In Kunzendorf RG (ed), Mental imagery. Plenum, New York, pp 45–52
9. Miller AI (1996) Insights of genius imagery and creativity in science and art. MIT Press, Cambridge
10. Winner E, Pariser D (1985) Giftedness in the visual arts. Items 31:65–69
11. Blazhenkova O, Kozhevnikov M (2009) The new object-spatial-verbal cognitive style model: theory and measurement. Appl Cognit Psychol 23:638–663
12. Kozhevnikov M, Blazhenkova O, Becker M (2010) Trade-off in object versus spatial visualization abilities: restriction in the development of visual-processing resources. Psychon Bull Rev 17:29–35
13. Kozhevnikov M, Kosslyn S, Shephard J (2005) Spatial versus object visualizers: a new characterization of visual cognitive style. Mem Cognit 33:710–726
14. Aspinwall L, Shaw KL, Presmeg N (1997) Uncontrollable mental imagery: graphical connections between a function and its derivative. Educational Stud Mathematics 33:301–317
15. Twyman A, (1972) Imagery and mathematical activity. The psychology of mathematics education. Chelsea College, London
16. Brewer W, Schommer-Aikins M (2006) . Scientists are not deficient in mental imagery: galton revised. Rev Gen Psychol, 10:104–120
17. Galton F (1880) Statistics of mental imagery. Mind 5:301–318
18. Kozhevnikov M, Hegarty M, Mayer RE (2002) Revising the visualizer-verbalizer dimension: evidence for two types of visualizers. Cognit Instruct 20:47–78
19. Jonassen DH, Grabowski BL (1993) Handbook of individual differences, learning, and instruction. Erlbaum, England
20. Kirby JR, Moore PJ, Schofield NJ (1988) Verbal and visual learning styles. Contemporary Educ Psychol 13:169–184
21. Denzin NK, Lincoln YS (2003) The landscape of qualitative research: theories and issues, 2nd edn. Sage, Thousand Oaks
22. Mays N, Pope C (2000) Qualitative research in health care. Assessing quality in qualitative research. Br Med J 320:50–52
23. Spradley JP (1994) The ethnographic interview. Holt, New York
24. Helitzer-Allen D, Makhambera M, Wangel A (1994) Obtaining sensitive information: the need for more than focus groups. Reprod Health Matters 3:75–82
25. Ulin PR, Robinson ET, Tolley E (2005) Qualitative methods in public health: a field guide for applied research. Family Health International and Jossey-Bas, San Francisco
26. Goering PN, Streiner DL (1996) Reconcilable differences: the marriage of qualitative and quantitative methods. Can J Psychiatry 41:491–497
27. Bauman LJ, Adair EG (1992) The use of ethnographic interviewing to inform questionnaire construction. Health Educ Q 19:9–23
28. Sheon N, Crosby MG (2004) Ambivalent tales of HIV disclosure in San Francisco. Soc Sci Med 58:2105–2118
29. Koester KA, Maiorana A, Vernon K et al. (2005) HIV surveillance in theory and practice: assessing the acceptability of California's non-name HIV surveillance regulations. Health Policy.
30. Patton MQ (1990) Qualitative evaluation and research methods, 2nd edn. Sage, Newbury Park
31. Patton MQ (2002) Qualitative evaluation and research methods, 3rd edn. Sage, Thousand Oaks

Ubiquitous Serendipity: Potential Visual Design Stimuli are Everywhere

Gabriela Goldschmidt

Stimuli as Inspiration

Designing is a purposeful activity; one engages in design with a goal in mind. The goal may be explicitly and fully articulated at the outset, in which case the designer is solving a well-defined problem, or it may be vague and somewhat ambiguous, in which case the problem is ill-defined. Furthermore, the problem may be more or less well (or ill-) structured, depending on whether or not there are established procedures for solving the problem and unequivocal criteria to evaluate the outcome. Many design problems, especially in architecture, industrial design and visual communication are notoriously ill-defined and ill-structured [2–4]. Moreover, in these fields it is all but a standard requirement to come up with the most innovative and creative solution possible. There are exceptions, of course: for example, when dealing with issues of safety in buildings or products creativity is not the first priority. But this is an exception that confirms the rule.

When faced with a new ill-defined/structured design assignment, then, a designer does of course try to avail him or herself of all the relevant data and information pertaining to the task, but the creativity imperative also forces one to look for a major idea with which to frame the design task [5]. We postulate that requirements, information, data and design methods—in combination or apart—do not directly translate into design ideas, which must be sought separately, while keeping all of the above in mind. Thus a search begins for design ideas (e.g., [6, 7]). The search is dynamic, because the ill-defined state of the problem allows for variations and tradeoffs in priorities and even in the conceptualization and interpretation of the problem itself and expectations from the solution. As we shall see, this is an important characteristic of designing with consequences for our current topic.

G. Goldschmidt (✉)
Israel Institute of Technology, Haifa, Israel
e-mail: gabig@tx.technion.ac.il

Center for Design Research, Stanford University, Stanford, USA

J. S. Gero (ed.), *Studying Visual and Spatial Reasoning for Design Creativity*,
DOI 10.1007/978-94-017-9297-4_12

A search for ideas occurs in a design space [8] that is as rich as the items that populate it are. These items, installed there by the designer, are in fact resources the designer can tap and use, among others, as sources of inspiration. Very often the items in the search space are visual images of one kind or another. Images may be retrieved from memory or summoned from within-domain or between-domain sources. The designer may deliberately choose them or simply invite them on board when stumbling upon them on the fly. They may also be presented to the designer by others, intentionally or unintentionally. The point is that visual images are often useful to the designer who hopes to be able to read into them information that may be helpful in developing design ideas. There is evidence that visual stimuli are inspirational and enhance design performance and in certain disciplines, like fashion design, designers routinely prepare 'inspiration boards' in which they display images that they wish to make use of in their designs [9]. Creativity or innovation scores assigned by judges are higher when designers are exposed to stimuli as compared to control groups that perform the same tasks with no stimuli (e.g., [10]). On the other hand exposure to examples of solutions to similar design problems may cause fixation, lowering the designer's capacity to be innovative (see below). However, even in such extreme cases it is not impossible for designers to circumvent the fixation problem, e.g., by re-defining the problem [11].

Gabora's Theory of Memory Structure and Creative Processes

To understand why images are potent sources of inspiration we should look into some of the characteristics of creative processes, at the cognitive level. To do so we shall build on Gabora [1], who proposes a theory of cognitive mechanisms underlying the creative process in terms of the structure and dynamics of memory processes. Gabora's departure point is the hypothesis that "creativity involves the capacity to spontaneously shift back and forth between analytic and associative modes of thought according to the situation." ([1], 2). In a creative assignment which necessitates an intuitive search, the first phase involves primarily associative, divergent thinking [12–14]. Thought is triggered by stimuli, which activate memory in specific patterns that support divergence or convergence. This activation is discussed below.

One of memory's important traits is that it is distributed, such that items are stored in various neurons, or 'memory locations' [15, 16]. Each instance of experience activates a large number of neurons, in a large yet constrained region. The pattern of activity distribution depends on its intensity, which is a function of stimulus frequencies. A high degree of activation causes more memory locations to be impacted around the most activated location. This is referred to as a flat activation function. In a low activation pattern we observe a spiky activation function, which means that a smaller and non-continuous number of memory locations are impacted, creating a spiky pattern around the most activated location. A flat activation function corresponds to divergent thought, because it includes a wider range of memory locations which are all at least somewhat different from one another. A spiky activation pattern corresponds to more analytic, convergent thought, because it involves only clusters of neurons that are very similar to each other in the contents they store.

Another important trait of memory is that it is content addressable. This means that different memory locations have specific 'addresses' and the content of an experience has a systematic relationship with the location where it is stored. This is also the location from which this content is retrievable. Gabora [1] explains that therefore, contents with related meaning are stored in overlapping locations and the address of a memory location is in fact determined by the pattern of activation that lead it to be affected. This is important because if more aspects of a stimulus are paid attention to and are encoded, more overlapping memory locations will be activated, resulting in more retrieval opportunities.

Various stimuli aspects may be paid attention to if one's attention to stimuli is defocused. There is some evidence that creativity is associated with defocused attention [17–19]: exerting defocused attention means perceiving more traits of a situation or stimulus. This results in affecting more memory locations that overlap and reinforce one another, and the activation function is therefore flat. Therefore there is more potential for associations among memories to be formed, resulting in what Gabora [1] refers to as associative richness. Heightened sensitivity of individuals, particularly to details, goes hand-in-hand with defocused attention, to affect a flat activation function and create potential for hitting stored memories of concepts, experiences or episodes that would not usually be associated with the stimulus that evoked them.

Memory activation is a dynamic process. When neurons in one region are activated, what is stored in those neurons triggers further activation. The thought is now somewhat different, and a slightly different region becomes activated. A cycle like this is responsible for a stream of thought. Streams of thought are necessary to create conceptual fluidity.

As we see, mechanisms of memory activation are not fixed and correspond to an individual's sensitivity and the way in which stimuli are perceived. We have the ability to exert control over our focus of attention, moving from focused to defocused attention and vice versa, thereby activating flat or spiky patterns of neuron activation. This makes it possible to switch back and forth between associative and analytic thinking which, according to Gabora [1], is typical of creative thinking. We must now relate this memory structure and its relationship with activity to what we know about the way stimuli are used by designers.

Creative Design Processes in Light of Gabora's Theory

There is enough empirical evidence to show that exposure to visual stimuli may enhance creative design problem solving [10, 20–23]. Visual stimuli may be carefully selected displays, or random images one comes across; even one's own sketches may become stimuli—designers infer unexpected information even from their own rudimentary sketches [24–26]. Other, non-visual types of stimuli seem to also have a positive effect on the rated creativity of design outcomes (e.g., [27] regarding textual stimuli). At the same time, exposure to straightforward solution examples at the outset of designing appears to have mostly negative effects, in the sense that designers are unable to divorce themselves from features of the example shown to them, and carry them over to their solution even when explicitly instructed not to do so [28–32].

How can we explain this based on Gabora's [1] cognitive theory of memory activation? In the case of examples taken from the same domain and the same problem, wherein designers are explicitly told not to repeat the stimulus object but rather come up with something new, one pays attention to stimulus features rather than the ensemble. These features, or traits, are perceived individually and given the content addressability of memory locations, they register there, provoking similar individual traits stored in the same neurons. Attention remains focused, producing a spiky activation pattern, and a limited number of locations in the region participate in the process. This amounts to convergent thinking, with little opportunity to break out and spread to more memory regions.

In contrast, when a stimulus is not directly related to the problem at hand, and better yet if it does not belong in the same field or discipline, one is 'invited' to perceive not just, or not primarily, individual traits. The totality of the stimulus is perceived and interpreted based on shape or form, or on specific contents and meaning it may suggest to the problem solver. This automatically expands the potential of different memory locations to be evoked because the problem solver may consider different connotations, or combinations of individual features, and in particular relationships among them. Attention is thus defocused and activation is spread wider, in a flat function, to more contents addresses. More overlaps occur and therefore there is potential for further new readings of memory locations, leading to wider associative and divergent thinking. I would like to postulate that perceiving totalities with less regard to details and interpreting them freely based on associations to what is already stored in memory is different from perceiving details. In repeating such processes one can gradually distance oneself from the original percept through cycles of associative thoughts, which yields, or may yield, what we call abstraction in thought. Oftentimes analogical reasoning is involved. The effect is particularly strong in between-domain analogies, where the stimulus pertains to a domain different from the one the problem belongs in [33]. Abstraction is a prerequisite to enhancing creativity through the use of stimuli in such reasoning processes [34]. Since the dynamic nature of the design search allows for modifications in goals and expectations, abstraction becomes easier because there are more directions in which associative thinking may venture to.

Another prerequisite for the successful use of stimuli in design, that is, one that enhances creativity, is that the designer be able to transform features of the stimulus rather than apply them more or less as is [34]. Transformation as opposed to simple transfer can also be explained with Gabora's [1] theory. Several researchers proposed that a certain amount of randomness in a problem solving process is

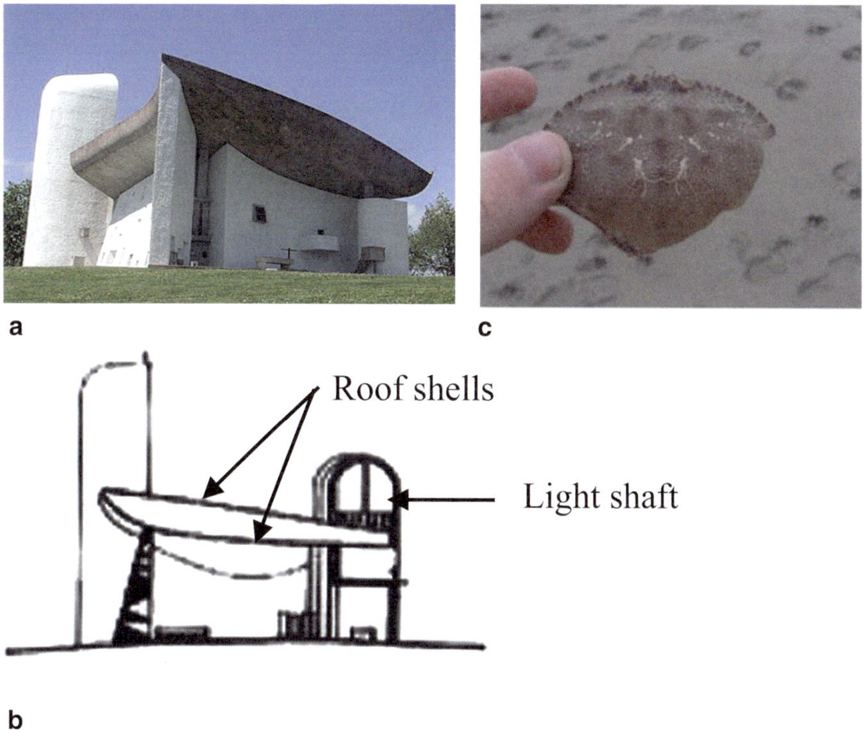

Fig. 1 Le Corbusier's Nôtre Dame du Haut chapel in Ronchamp, 1955—roof inspired by crab shell. **a** Photo. **b** Section drawing. **c** Crab shell

conducive to, or even indispensible for, creative outcomes [13, 35]. In a search in the early phases of design problem solving a designer may seek inspiration in a non-structured way, that is, simply pay attention to random stimuli around him or her, hoping that something will trigger a constructive idea related to the problem he or she is out to solve. If this occurs, one experiences an 'Aha!' moment. This is not unique to design of course; similar experiences are reported for example in scientific breakthroughs. Anecdotal evidence can be found in the analogy literature and the mental imagery literature, as well as in the design literature. Consider for example Le Corbusier's Nôtre Dame du Haut chapel in Ronchamp (Fig. 1). According to his own testimony Le Corbusier used as stimulus a crab shell he had on his drawing board, picked up on a beach a few years earlier. He noticed the strength of the shell despite its minimal thickness, which is attributable to its geometric form of two curvilinear surfaces joined at the edges. This led to the design of the chapel roof that is composed of two shells of reinforced concrete only 2 cm thick, 2 m apart in the widest place, and joined at the edges. The stimulus was thus a remote between-domain source, having nothing to do with the chapel Le Corbusier was designing, and which needed to undergo a serious transformation in order to contribute to the chapel's design. This kind of thinking is definitely defocused, wherein

Le Corbusier's attention wandered between his building and the natural phenomenon by which he had been enchanted. There could not have been a single memory address where similar items were stored, so most likely a number of locations with some kind of relationship to crabs and to thin membranes were activated, to create a diffused memory distribution and flat activation function. Associating, divergent thinking is obvious in this case.

However, this was not the only stimulus Le Corbusier is said to have used in the design of the Ronchamp chapel. It is claimed he consciously, and not randomly, "borrowed" from Adrian's villa in Pompeii as well as other architectural sources for the design of the light shafts and the battered walls of the chapel. Those stimuli are within-domain, being buildings, and here Le Corbusier's thinking was focused and convergent; direct paths led to memory locations which stored light shafts and battered walls which Le Corbusier had seen and been impressed with, and remembered well.

In the design of the Ronchamp chapel, then, Le Corbusier had most probably boasted flexible thinking in which he shifted from focused to defocused thinking and vice versa in a fluid mode until the combination of the different elements afforded by these modes of thinking came together to create an integrated design of the entire building and its various components.

Le Corbusier and others of similar stature are not the norm, but many designers (and other problem solvers) go through similar thought processes, even though the resulting designs or discoveries may be of lesser historical dimensions. Why is it that some designers make creative breakthroughs and others do not? In other words, what does it take to be able to activate the kind of cognitive processes that lead to creative thinking?

The Prepared Mind

Randomness in a design search may lead to an opportune encounter with stimuli that the designer is able to tie to unexpected items in memory, through a chain of associations, thus paving the way to a creative design idea. This may happen by chance; however, it does not always happen and it does not happen to everyone: as the saying goes, "chance favors the prepared mind." What is a prepared mind in the context of design?

Gabora [1] has laid out two general prerequisites for a creative search. First, as already quoted earlier, having "the capacity to spontaneously shift back and forth between analytic and associative modes of thought". The second prerequisite is possessing heightened sensitivity. Sensitivity, however, cannot be expected to be identical across the board and the subject matter to which one is sensitized to may differ between domains. This has implications for the items we store in memory: we do better with items directly related to our interests and expertise. For example, Chase and Simon [36] have shown that master chess players remember an unusually large number of game boards, whereas their memory is no more than average for other matters. Kokotovich and Purcell [37] found that designers outperform non-designers in creative mental synthesis tasks (involving shape and form combinations). Moreover, they tested two groups of designers—visual communications designers who normally deal with 2D shapes, and industrial designers who routinely manipulate 3D forms. They found differences between the design disciplines in their output where superior performance occurs when the type of parts to be synthesized are similar to the parts normally used to produce shapes or forms within the discipline. We may therefore conclude that in order for a mind to be prepared, its owner should also have a certain amount of expertise in the domain or sub-domain in which he or she operates.

Since stimuli in design are for the most part visual, it is natural to expect the designer, and certainly the creative designer, to have the benefit of high-level visual literacy. Wikipedia[1] defines visual literacy as follows: "Visual literacy is the ability to interpret, negotiate, and make meaning from information presented in the form of an image. Visual literacy is based on the idea that pictures can be "read" and that meaning can be communicated through a process of reading." People with good visual literacy are able to make meaningful interpretations of visual stimuli that mean little or nothing to those with low visual literacy. With good visual literacy comes the ability to infer information from an image, and as we have seen in design assignments, even unexpected and surprising information in one's own sketches that may be crucial to success in the case of problem solving [24–26].

We propose that the creative designer's 'prepared eye' allows him or her to 'hunt' for helpful clues in whatever stimuli he or she comes across, intentionally or randomly; the 'hunting' mode is almost an instinct which is always switched on. 'Hunting' may take place in two modes: First, when it is to serve a particular design search in which a 'good' stimulus, if found, that is, one boasting a 'good-fit' [38], may enhance creativity in the way we described above. But there is also a constant 'hunt' that creative designers are engaged in, not necessarily related to a particular design assignment, but rather for the purpose of building up an inventory of potentially useful images. When these are stored in memory one is able to tap them at the appropriate time, when activated by a currently perceived stimulus. There is plenty of anecdotal evidence to show that designers go on 'visual hunts'. For example, the architect Denys Lasdun [39] wrote:

> In the course of creation an architect may receive inspiration from a large number of sources from works of the past and the present and from right outside architecture. He must have something to work on; he is certainly no less creative if he spreads his net wide and has an eye that remembers. (p. 107).

Le Corbusier, whose use of stimuli in the design of the Ronchamps chapel we described above, was an avid collector of images [40]:

> His [Le Corbusier's] mind was well stocked with ideas, devices, configurations and images gleaned form tradition, from painting, from observation, and of course from his own earlier works… At the right moment images would flow to the surface where they would be caught, condensed and exteriorized as sketches. (11).

[1] Wikipedia, accessed March 20, 2010.

Keller [41] has recently proposed a digital equivalent of the traditional 'Cabinet' in which cherished objects of inspiration are stored for future reference. The architect James Stirling talked about his tendency to 'appropriate' images he came across and liked, in order to use them in his design work. He referred to himself as a 'magpie' that does not shy away from 'stealing' images [42]:

> Like Picasso, Stirling operated a magpie avidity to steal whatever he liked while yet turning it into his own… (20).

The evidence suggests that for the creative designer who is an expert in his domain, possesses an 'educated eye' acquired through a high level of visual literacy, and whose sensitivities are geared towards a never ending scrutiny for 'treasures' that can be stored in memory or used on the spot—the world is a warehouse of potential stimuli. Stimuli are ubiquitous if one is attentive enough to be able to capture and harness them so as to serve as sources of inspiration in design. Moreover, the experienced treasure hunter with a good eye will encounter serendipity, as he or she learns to single out stimuli with the highest amount of potential. Similar to expert chess players who can anticipate a long chain of moves in a chess game, the expert designer can anticipate the usability of a stimulus image through the chain of memory activation described above. His or her mind is well stocked with memory items that have either been helpful in the past or have impressed the designer as holding promise to be useful in the future. In addition, stimuli will be chosen for both their totality and their components, and our designer is able to switch attention from one to the other, activating associative and analytic thinking as per need. 'Ubiquitous serendipity' is therefore not an exaggerated way to describe the way in which designers capitalize on cognitive mechanisms to exploit stimuli everywhere and any time, to enhance their design creativity.

Conclusions

Two major factors are involved in creative processes, and in particular in designing with stimuli: The brain's architecture controlling memory processes, and the designer's preparedness. Memory activation functions are only minimally controllable, in focusing and defocusing attention to stimuli, in their totalities and/or details. Preparedness requires that the designer be adequately expert in his or her domain, and that he or she possess a high-level of visual literacy, since most stimuli that have been shown to be effective in enhancing design creativity are visual. Having a large inventory of stored images acquired over time and ready to be tapped as per need is therefore also highly relevant. Being endowed with heightened sensitivity is indispensible and this means that the designer is always alert to take in new stimuli as they present themselves, by chance or otherwise. I believe this explains creativity in design and could possibly serve to develop tests which may be able to predict who would make a creative designer at the outset of design education.

Acknowledgement The writing of this paper was partially supported by a grant to the author from the fund for the promotion of research at the Technion, hereby gratefully acknowledged. The author would also like to thank Stanford University's CDR for hosting her while on sabbatical and providing a stimulating intellectual environment at the time this paper was written.

References

1. Gabora L (2010) Revenge of the 'neurds': characterizing creative thought in terms of the structure and dynamics of memory. Creativity Res J 22(1):1–13
2. Goel V, Pirolli P (1992) The structure of design problem spaces. Cognitive Sci 16:345–429
3. Lynch C, Ashley K, Aleven V, Pinkwar N (2006) Defining Ill-Defined Domains; a literature survey. In: Aleven V, Ashley K, Lynch C, Pinkwart N (eds) Proceedings of the Workshop on Intelligent Tutoring Systems for Ill-Defined Domains at the 8th International Conference on Intelligent Tutoring Systems. Jhongli (Taiwan), National Central University, pp 1–10
4. Simon HA (1973) The structure of ill-structured problems. AI 4:181–201
5. Darke J (1979) The primary generator and the design process. Des Stud 1(1):36–44
6. Cross N (2006) Designerly ways of knowing. Springer, London
7. Lawson B (2006) How designers think, (4th edn.). Elsevier, Oxford
8. Woodbury RF, Burrow AL (2006) Whither design space? AIEDAM 20(2):63–82
9. Eckert CM, Stacey MK (2000) Sources of inspiration: a language of design. Des Stud 21(5):523–538
10. Casakin H, Goldschmidt G (2000) Reasoning by visual analogy in design problem-solving: the role of guidance. J Plan Des: Environ Plan B 27:105–119
11. Goldschmidt G, Ben-Zeev A, Levi S (1996) Design problem solving: the effect of problem formulation on the solution space. In Proceedings of the Thirteenth EMCSR '96, Vienna, pp 388–393
12. Boden M (1990) The creative mind: myths and mechanisms. Weidenfeld & Nicolson, London
13. Johnson-Laird PN (1983) Mental models. Harvard University, Cambridge
14. Sloman S (1996) The empirical case for two systems of reasoning. Psychol Bull 9(1):3–22
15. Hinton G, McClelland JL, Rummelhart DE (1986) Distributed representations. In: Rummelhart DE, McClelland JL (eds) Parallel distributed processing: explorations in the microstructure of cognition. MIT, Cambridge
16. Kanerva P (1988) Sparse distributed memory. MIT, Cambridge
17. Martindale C (1999) Biological bases of creativity. In: Sternberg R (ed.) Handbook of creativity. Cambridge University, Cambridge, pp 137–152
18. Mednick SA (1962) The associative basis of the creative process. Psychol Rev 69:220–232
19. Mendelsohn GA (1976) Associative and attentional processes in creative performance. J Pers 44:341–369
20. Bonnardel N (2000) Towards understanding and supporting creativity in design: analogies in constrained cognitive environments. Knowl Based Syst 13:505–513
21. Goldschmidt G (2001) Visual analogy-a strategy for design reasoning and learning. In: Eastman C, Newsletter W, McCracken M (eds) Design knowing and learning: cognition in design education. Elsevier, New York, pp 199–219
22. Goldschmidt G, Smolkov M (2006) Variances in the impact of visual stimuli on design problem-solving performance. Des Stud 27(5):549–569
23. Verstijnen IM, Hennessey JM, Leeuwen C, Hamel R, Goldshmidt G (1998) Sketching and creative discovery. Des Stud 19(4):519–546
24. Goldschmidt G (1991) The dialectics of sketching. Creativity Res J 4(2):123–143
25. Suwa M, Tversky B (1997) What do architects and students perceive in their design sketches? A protocol analysis. Des Stud 18(4):385–403

26. Do EY-L, Gross MD (1995) Drawing analogies: finding visual references by sketching. In Proceedings of Association of Computer Aided Design In Architecture (ACADIA), Seattle, pp 35–52
27. Goldschmidt G, Litan A (2009) From text to design solution: Inspiring design ideas with texts. In Proceedings of ICED'09, August 24–27, Stanford University, Palo Alto, pp 9.15–9.26
28. Cardoso C, Badke-Schaub P (2009) Give design a break? The role of incubation periods during idea generation, vol 2. In Proceedings of ICED'09, Stanford University, pp 383–394
29. Jansson DG, Smith SM (1991) Design fixation. Des Stud 12(1):3–11
30. Linsey J, Tseng I, Fu K, Cagan J, Wood KL (2009) Reducing and perceiving design fixation: Initial results from an NSF-sponsored workshop in Proceedings of ICED'09, vol 9. Stanford University, pp 233–244
31. Perttula MK, Liikkanen LA (2006) Exposure effects in design idea generation: Unconscious conformity or a product sampling probability? In: Jonsson M, Unnporson R (eds) Proceedings of NordDesign 2006, University of Iceland, Reykjavik, pp 42–55
32. Purcell AT, Gero JS (1996) Design and other types of fixation. Des Stud 17(4):363–383
33. Bonnardel N, Rech M (1998) Les sources d'inspiration en conception. Sciences et Techniques de la Conception 6:37–53
34. Goldschmidt G (2010) Avoiding design fixation: Transformation and abstraction in mapping from source to target. In: Cardoso C, Badke-Schaub P (eds) Proceedings of International workshop on: Fixation or inspiration? The role of internal and external sources on idea generation, TUDelft, April 12–13, 2010 (no pp)
35. Dennett D (1978) Brainstorms: philosophical essays on mind and psychology. Harvester Press, Brighton
36. Chase WG, Simon HA (1973) The mind's eye in chess. In: Chase WG (ed) Visual information processing. Academic, New York, pp 215–281
37. Kokotovich V, Purcell T (2000) Mental synthesis and creativity—an experimental examination. Des Stud 21(5):437–449
38. Alexander C (1964) Notes on the synthesis of form. Harvard University, Cambridge
39. Lasdun D (1976) A language and a theme: the architecture of Denys Lasdun & partners. RIBA Publications Ltd, London
40. Curtis WJR (1986) Le Corbusier: ideas and forms. Rizzoli
41. Keller I, Visser FS, van der Lugt R, Jan P (2009) Collecting with cabinet, or how designers organise visual material, researched through an experimental prototype. Des Stud (1):69–86
42. Wilson C St. John (1992) James Stirling: in memoriam. Architectural Rev 191(1150):18–20

On Abstraction and Ambiguity

Barbara Tversky

Spatial thinking ranges from the concrete—moving through space, moving things in space—to the abstract—moving through conceptual spaces, moving ideas in conceptual spaces. Both concrete and abstract spaces can be frozen and externalized in objects, maps, and diagrams, which in turn can be moved through and manipulated, in the mind or in the world. Creating representations of space, real or metaphoric, in the mind or outside of it, requires abstraction, and consequent ambiguity. Ambiguity enables the multiplicity of interpretations that are the foundation of creative thought. Space, action, and abstraction are inextricably interlinked: spraction.

Internal Spaces

Mental Representations

We inhabit many spaces. There is the space of our own bodies. There is the space immediately around our bodies, within grasp of eye or hand. There are the larger spaces, real or imagined, too large to be seen at a glance. Spaces that we may move about in, spaces we may arrange and rearrange, either in body or in mind. There are the spaces we create, the spaces we live and work in, the spaces we put on paper to augment our minds and to communicate, to self or other. Each space connects perception, thought, and action. Mental spaces are constructed out of what we perceive, aided by what we think and infer, in the service of action, in the world or imagined in the mind. Corresponding to the multiplicity of spaces is a multiplicity of kinds of representations of space.

B. Tversky (✉)
Stanford University, Stanford, USA
e-mail: bt@psych.stanford.edu

Columbia Teachers College, Columbia University, New York, USA

J. S. Gero (ed.), *Studying Visual and Spatial Reasoning for Design Creativity*,
DOI 10.1007/978-94-017-9297-4_13

As such, mental representations of space are not constructed from a single experience nor from a single modality. We experience space from particular viewpoints at particular times, yet place and time are constantly changing. We must have more general representations to accommodate to those changes. Laboratory rats, and undoubtedly a multitude of other creatures, form general representations of space online as they move about new environments [37]. We know about space from sight, from sound, from smell, from the kinesthetic and proprioceptive sensations from our own bodies. Each modality adds something unique, depending on the task, but to some extent, one can substitute for another. Many blind people navigate remarkably well, and many are remarkable spatial thinkers.

Nor are mental representations stored in a mental bookshelf like paper maps, and pulled out and looked at when needed. Rather, mental representations are more analogous to Google maps, created and recreated on demand to suit users' changing needs. The mind is, among other things, a vast data base, from which specific information can be retrieved and organized to suit current purposes. We establish mental representations as and when we need them, drawing on whatever information that seems relevant that we can find.

Useful mental representations, then, are constructed from multiple experiences and multiple modalities. That entails integrating those different experiences and different sources. How might that integration be done? Consider, for example, the larger spaces that we navigate. We learn about spaces that we navigate from the many views of actual navigation, from maps, from descriptions. Various kinds of evidence suggest that the mind attempts to align these different experiences and modalities the way a surveyor or a geometer might, by selecting shared elements, identifying a reference frame for the elements, and aligning the reference frames and elements. Each of these processes, of course, entails subprocesses, finding the critical elements, determining the appropriate reference frame, and then aligning them. Unlike good surveyors or geometers, the mind does this alignment crudely, approximately, not with numbers and exact calculations. Indeed, distortions in spatial memory arise from simplifying the the relations among elements and the relations between elements and reference frames, suggesting that both serve as anchors for constructing mental representations of things in space [50, 52]. Specifically, large elements, like North and South America, are remembered as more aligned with each other than they actually are, and large elements, like South America, are remembered as more aligned with a reference frame than they actually are.

Thus for the space of navigation, mental representations are constructed from key geographic elements, by relating them to each other and to an encompassing frame of reference. Objects and reference frames may differ for different spaces. For the space around the body, the reference frame is extensions of the axes of the body, front/back, head/feet, left/right [11]. For the space of the body, the configuration of the parts of the body form a reference frame, with the body parts as elements, but where their functional significance overshadows their actual size [31].

Mental representations, then, are abstractions. They combine, summarize, stand for many different experiences and many different modalities. The construction processes that create them by their very nature simplify and inevitably distort the information. Spatial representations are not alone in this abstraction process. Language, too, abstracts, simplifies, distorts. One important consequence of abstraction, spatial or verbal, is ambiguity. Just as a representation is a summary of many instances, a many-to-one mapping, a representation has many possible interpretations, a one-to-many mapping. Finding new interpretations is a key process in creativity, in science, in art, in play, and in design. And key to new interpretations are mental transformations.

We have now established the first link in the argument we are building: the link from the world to the mind through abstraction, and its shadow, ambiguity. Yet mental representations are useless without thought, without using them in one way or another. Using mental representations entails transformations of them.

Mental Transformations

Mental representations are not usually useful in and of themselves. Even pictures on the wall need interpretation or evaluation. Mental representations become useful when they are used, for making inferences and judgments. A mental representation of an environment for example, can be used to find a route, to estimate distances, to plan new buildings, to arrange new streets. Inferences and judgments on mental or external representations require mental or actual transformations of the representations. One kind of mental transformation, renown even in the public at large, is mental rotation [43]. That task entails deciding whether two very similar meaningless, complex, three-dimensional figures depicted at different angles are the same or mirror images of each other. For highly practiced participants, the time to make the judgment increases linearly with the angle separating the two figures [44]. Performance on this task correlates with a remarkable number of spatial reasoning tasks [21]. For example, it correlates with how quickly and accurately participants assemble a new piece of furniture and it correlates with the quality of both their visual and their verbal explanations of this task [8, 60]. However, it does not correlate with other visuospatial reasoning tasks, some key to design, notably, finding new interpretations in ambiguous sketches. Another kind of mental transformation does predict finding new interpretations in ambiguous sketches, specifically, a task requiring close attention to details of visual forms and abstraction, detecting geometric forms embedded in complex figures [46].

Just as there are many ways to manipulate objects and spaces in the world, there are many ways to manipulate them in the mind. There are a multitude of possible mental transformations [52]. In addition to changing orientation of figures, people can imagine changes in locations, changes of size or shape, additions or subtractions of parts, reconfigurations, rearrangements, and more. People can mentally count or estimate length, width, area, proportion, direction, speed, and relative quantity, length, area, direction, speed, as well as changes in and combinations of these. The multitude of mental transformations the mind can perform are not merely useful but essential as much for our daily lives as for our leaps of creativity.

Intriguingly, many of these mental transformations appear to be internalizations of physical transformations, either viewed or performed or both. An illustrative example comes from solving geometric analogies [36]. Transformations like change location, change orientation, add or subtract parts, cut in half or double symmetrically are the basis for problems in geometric analogies. When participants are asked to solve geometric analogies that require two or three such transformations, people perform the transformations in a stereotyped order, even though the order makes no difference to the correctness of the solution. When they are asked to solve the problems using a different order, they are slower and make more errors. The order, surprisingly, does not depend on spatial working memory demands, as the order does not correspond to either difficulty or ease of the transformation. Rather, the order of performing the mental transformations in analogy solution corresponds to drawing order, as if participants had internalized drawing operations and were mentally drawing the solutions.

The claim that mental transformations are internalizations of physical transformations is strengthened by recent work on brain processing on the one hand and on gesture on the other. Surprisingly, motor or premotor cortices are involved in mental transformations such as mental rotation [10, 11] and counting [1, 27], even when no actual motor activity is involved.

Mirroring the work implicating motor cortex in mental transformations are findings that actual movements, when congruent with mental transformations, promote mental transformations. Mental rotation is facilitated by congruent rotations of the hands [7, 61, 62], as is mental counting [4]. Solving problems that entail spatial transformations such as the water level problem [41] or the directions of gears [42] is facilitated by congruent gestures. These findings and others like it strongly implicate external as well as internalized action as a basis for spatial mental transformations. Thus we have established another essential link in the argument we are developing: a link between thought and action, between mental transformations and physical ones.

External Spaces

Even before the advent of Homo sapiens, creatures created artifacts and arranged environments to better their lives. Nevertheless, the built world that we inhabit attests to the scale in which humans engage in these activities. Humans are perhaps unique in creating artifacts that improve not just their physical but their mental lives, notches in sticks, incisions in stone, paintings in caves, maps on paper, diagrams in textbooks, virtual worlds in bits. These myriad cognitive tools augment cognition in myriad ways, among them: they expand memory, both long-term and working, they facilitate information processing, they depict ideas so others can understand them and collaborate in their revision [9, 35, 52]. Importantly, they abstract thought, in much the way that internal mental representations do.

Maps are an ancient and prototypic example. They are an admirable feat of the mind. From experiencing space from within in, we are able to integrate and abstract what is essential, typically, the configuration of landmarks and paths, to imagine how that configuration would look from above, and to shrink it, so as to inscribe it on a stone or print it on paper and fold it into a pocket.

In capturing landmarks and paths, maps omit large quantities of information in environments, buildings, trees, and more, information that is not needed for the usual uses of maps. Maps, of course, are not uniform, they include, exclude, minimize, exaggerate, and choose perspectives as appropriate to their anticipated uses, among them, to determine routes, to estimate directions and distances, to understand processes that unfold in space, from traffic of cars to traffic of pollen, populations, and pollution. Highway maps exaggerate the size of roads and rivers; otherwise they would not be visible at typical scales. Tourist maps mix perspectives, showing the network of roads as if from above but the tourist attractions as if from within, so that tourists can recognize the landmarks on the ground and use them for navigation.

Maps are privileged. As communications, they convey elements and relations in in the space of the world by using elements and relations in the space on paper. This mapping to space confers an enormous advantage over symbolic representations, an advantage maps share with other forms of visual communication, diagrams, charts, graphs, gesture–visualizations of things that are not inherently visible. Maps use elements and spatial relations on paper to convey elements and spatial relations in the world. Diagrams use elements and spatial relations on paper to map metaphorically spatial ideas and relations [51]. Gestures perform feats similar to those of maps and diagrams, using movements of the body, especially the hands, to enact a virtual space, locate things in it, and often enact change in virtual elements and their relations. Gestures, of course, are part of being human, and even precede Homo sapiens [49]. Although diagrams proliferate today in newspapers, textbooks, instructions, and even billboards, compared to maps, which are ancient, diagrams, charts, and graphs are relatively new.

By spatializing abstract ideas, diagrams allow a rich set of spatial mental transformations to be applied to abstractions to enable inferences about abstractions [51]. It is not that our spatial inferencing abilities are perfect, they are certainly not, but that they are more highly developed than our abstract inferencing abilities, having served us from birth, if not before. No wonder the proliferation of visualizations and no wonder the ubiquity of gesture, they not only externalize thought but also promote inferences.

For many everyday uses, abstraction and clarity are desiderata. A good map should enable users to find routes easily, to make spatial inferences easy. Good maps should not be cluttered with detail that distract from those tasks; this is one reason why aerial photographs make poor maps. Think of sketch route maps and well-designed subway maps: they typically distort paths and directions to support way-finding [56, 57]. Similarly, good instructions for assembling or operating something or for understanding how things work emphasize the steps needed to assemble, operate, or work in part by eliminating extraneous features [60]. These visual communications tailor their abstraction to their uses to enhance performance and minimize error.

Yet, clarity is not always desirable. A common complaint about CAD-CAM programs is their clarity; they rectify a design that is not ready to be rectified, they specify details that are not ready to be specified. In freezing ideas, they can freeze thought. This is why designers often prefer sketching. Sketches look as ambiguous as the ideas they are meant to express. Often they show the parts that should be there, and even their locations, but indefinitely.

Significantly, it seems their very sketchiness, their ambiguity, is what makes sketches easier to think about, to contemplate, to find unintended discoveries, to see new possibilities, to revise and improve. These processes, iterations of sketching and contemplating, become a kind of conversation between designer and sketch, a positive cycle that has been documented [13, 18–20, 40, 46, 47, 54, 58]. Expert designers are better at seeing new interpretations in sketches than novices. Those adept at remote associations and/or those proficient at detecting hidden figures in complex ones are better at finding new interpretations [46]. These skills enable a pair of processes important to creative thought, reinterpreting the perception and finding meaning in the reinterpretation [46, 54, 58].

One strategy that expert architectures report using to find unintended discoveries and that designers report using to find new interpretations of sketches is reconfiguring the elements of the sketch. Reconfiguring is just one of the many transformations in our mental handbooks. It seems likely that other transformations will enable the flexible thinking underlying creative thought. These transformations are bottom-up visuospatial mental transformations; it is also likely that top-down conceptual strategies would promote new ideas and insights as well [5]. The interplay of visuospatial and conceptual processes in the creation and development of new ideas is a promising research direction.

Sketching, then, expresses thought and fosters innovative thinking. Sketches have their visual side, but they are created by actions in space, by gestures. Intriguingly, blind-folding architects as they designed did not decrease the quality of their designs [3]. This finding should probably not be taken to show that sketching is not needed for effective design. Blind-folded designers led to copious gesturing, actions in space that mirrored the design ideas under consideration. And creating an external representation, such a sketch, is nearly ubiquitous in design; there must be reasons. That is, designers, could always gesture instead of sketching, but they typically choose sketching. External aids are more permanent than gestures; as such they offload memory, freeing the mind to do other things, providing a concrete platform for considering possibilities.

Ambiguity and Abstraction

The links are complete. Mental representations of space (and everything else) abstract, they select what are meant to be the critical elements and relations, perhaps exaggerate them, and omit considerable detail regarded as irrelevant, that may clutter and distract. External representations do the same, but in addition, provide a visible platform for productive thought. External representations can convey concepts that are inherently spatial as well as those that are metaphorically spatial. Abstractions are inherently ambiguous; they represent many possible states of affairs. Their ambiguity allows the taking of new perspectives and new interpretations, the foundations of creative thought. Spatializing ideas, both those that are inherently visuospatial and those that are metaphorically visuospatial, allows a cornucopia of visuospatial mental transformations to enhance productive thinking. Making and revising representations, whether internal or external, reflect and internalize actions in space. Space, action and abstraction are inextricably intertwined, captured by a new concept: spraction.

Acknowledgements Gratitude to the following grants for partial support of some of the research: National Science Foundation Grants HHC 0905417, IIS-0725223, IIS-0855995, REC-0440103, Office of Naval Research Grants NOOO14-PP-1-O649, N00014011071, and N000140210534, and the Stanford Visual Analytics Center.

References

1. Andres M, Seron X, Olivier E (2007) Contribution of hand motor circuits to counting. J Cognit Neurosci 19:563–576
2. Alibali MW, Bassok M, Olseth KL, Syc SE, Goldin-Meadow S (1999) Illuminating mental representations through speech and gesture. Psychol Sci 10:327–333
3. Bilda Z, Gero JS, Purcell T (2006) To sketch or not to sketch? that is the question. Des Stud 27:587–613
4. Carlson RA, Avraamides MN, Cary M, Strasberg S (2007) What do the hands externalize in simple arithmetic? J Exp Psychol Learn Mem Cognit 33:747–756
5. Chrisykou EG, Weisberg RW (2005) Following the wrong footsteps: fixation effects from pictorial examples in a design problem-solving task. J Exp Psychol Learn Mem Cognit 31:1134–1148
6. Corter E, Rho YJ, Zahner D, Nickerson JV, Tversky B (2009) Bugs and biases: diagnosing misconceptions in the understanding of diagrams. *Proceedings of the 31st Annual Conference of the Cognitive Science Society*
7. Chu M, Kita S (2008) Spontaneous gestures during mental rotation tasks: insights into the microdevelopment of the motor strategy. J Exp Psychol Gen 137:706–723
8. Daniel M-P, Tversky B, Heiser J (submitted) How to put something together
9. Donald M (1991) Origins of the modern mind: three stages in the evolution of culture. Harvard University Press, Cambridge
10. Eisenegger C, Herwig U, Jancke L (2007) The involvement of primary motor cortex in mental rotation revealed by transcranial magnetic stimulation. Eur J Neurosci 4:1240–1244
11. Ganis G, Keenan JP, Kosslyn SM, Pascual-Leone A (2000) Transcranial magnetic stimulation of primary motor cortex affects mental rotation. Cereb Cortex 10:175–180
12. Goel V (1995) Sketches of thought. MIT Press, Cambridge
13. Goldin-Meadow S (2003), Hearing gesture: how our hands help us think. Harvard University Press, Cambridge
14. Goldin-Meadow S, Nusbaum H, Kelly SD, Wagner S (2001) Explaining math: gesturing lightens the load. Psychol Sci 12:516–522
15. Goldin-Meadow S, Wagner SM (2005) How our hands help us learn. Trends Cognit Sci 9:234–241

16. Graham TA (1999) The role of gesture in children's learning to count. J Exp Child Psychol 74:333–355
17. Goldschmidt G (1991) The dialectics of sketching. Creativity Res J 4:123–143
18. Goldschmidt G (1994) On visual design thinking: the vis kids of architecture. Des Stud 15:158–174
19. Gross MD (1996) The electronic Cocktail Napkin—a computational environment for working with design diagrams. Des Stud 17:53–69
20. Franklin N, Tversky B (1990) Searching imagined environments. J Exp Psychol Gen 119:63–76
21. Hegarty M, Waller D (2005). Individual differences in spatial abilities. In: Shah P, Miyake A (eds) The Cambridge handbook of visuospatial thinking. Harvard University Press, Cambridge, pp 121–169
22. Heiser J, Tversky B (2006) Arrows in comprehending and producing mechanical diagrams. Cognit Sci 30:581–592
23. Heiser J, Phan D, Agrawala M, Tversky B, Hanrahan P (2004) Identification and validation of cognitive design principles for automated generation of assembly instructions. *Proceedings of Advanced Visual Interfaces '04*, ACM: 311–319
24. Hoestteter AB, Alibali MW (2008) Visible embodiment: gestures as simulated action. Psychonomic Bull Rev 15:495–514
25. Hoestteter AB, Alibali MW, Kita S (2007) I see it in my hands' eye: representational gestures reflect conceptual demands. Lang Cognit Process 22:313–336
26. Iverson J, Goldin-Meadow S (1997) What's communication got to do with it? Gesture in children blind from birth. Dev Psychol 33:453–467
27. Kansaku K, Carver B, Johnson A, Matsuda K, Sadato N, Hallett M (2007) The role of the human ventral premotor cortex in counting successive stimuli. Exp Brain Res 178:339–350
28. Kessell AM, Tversky B (submitted) Thinking with hands and paper
29. Krauss RM, Chen Y, Gottesman RF (2000) Lexical gestures and lexical access: a process model. In: McNeill D (ed) Language and gesture. Harvard University Press, Cambridge, pp 261–283
30. McNeill D (1992) Hand and mind: what gestures reveal about thought. University of Chicago Press, Chicago
31. Morrison JB, Tversky B (2005) Bodies and their parts. Mem Cognit 33:696–709
32. Nickerson JV, Tversky B, Corter JV (2014) Showing connection. In: Gero JS (ed) Studying visual and spatial reasoning for design creativity. Springer, Berlin
33. Nickerson JV, Corter J, Tversky B, Zahner D, Rho YJ (2008a) Diagrams as a tool in the design of information systems. In: Gero JS, Goel A (eds) Design computing and cognition '08. Springer, Berlin
34. Nickerson JV, Corter J, Tversky B, Zahner D, Rho YJ (2008b) The spatial nature of thought: understanding systems design through diagrams. *Proceedings of the International Conference on Information Systems*, Paris
35. Norman DA (1993) Things that make us smart: defending human attributes in the age of the machine. Perseus, New York
36. Novick LR, Tversky B (1987) Cognitive constraints on ordering operations: the case of geometric analogies. J Exp Psychol Gen 116:50–67
37. O'Keefe J, Nadel L (1978). The hippocampus as a cognitive map. Oxford University Press, Oxford
38. Rizzolatti G, Arbib MA (1998) Language within our grasp. Trends Neurosci 27:169–172
39. Rizzolatti G, Craighero L (2004) The mirror neuron system. Annu Rev Neurosci 27:169–192
40. Schon DA: (1983). The reflective practitioner. Harper Collins, Hammersmith
41. Schwartz D (1999) Inferences through imagined actions: knowing through simulated doing. J Exp Psychol Learn Mem Cognit 25:116–136
42. Schwartz DL, Black JB (1996) Shuttling between depictive models and abstract rules: induction and fall-back. Cognit Sci 20:457–497

43. Shepard RN, Cooper L (1982) Mental images and their transformations. MIT Press, Cambridge
44. Shepard RN, Metzler J (1971) Mental rotation of three-dimensional objects. Sci 171:701–703
45. Suwa M, Tversky B (1997) What architects and students perceive in their sketches: a protocol analysis. Des Stud 18:385–403
46. Suwa M, Tversky B (2003) Constructive perception: A skill for coordinating perception and conception. In *Proceedings of the Cognitive Science Society Meetings*
47. Suwa M, Tversky B, Gero J, Purcell T (2001) Seeing into sketches: regrouping parts encourages new interpretations. In: Gero JS, Tversky B, Purcell T (eds) Visual and spatial reasoning in design. Key centre of design computing and cognition. University of Sydney press, Sydney, pp 207–219
48. Taylor HA, Tversky B (1992) Descriptions and depictions of environments. Mem Cognit 20:483–496
49. Tomasello M (2008) Origins of human communication. MIT Press, Cambridge
50. Tversky B (1981) Distortions in memory for maps. Cognit Psychol 13:407–433
51. Tversky B (2001) Spatial schemas in depictions. In: Gattis M (ed.) Spatial schemas and abstract thought. MIT Press, Cambridge
52. Tversky B (2005). Visuaospatial reasoning. In: Holyoak K, Morrison R (eds) The Cambridge handbook of thinking and reasoning. pp 209–241. Cambridge: Cambridge University
53. Tversky B (2011). Visualizations of thought. Top Cognit Sci 3:499–535
54. Tversky B (In press, a) Making thought visible. In: Gero J (ed) Studying design creativity. Springer, Berlin
55. Tversky B, Heiser J, Lee P, Daniel M-P (2009) Explanations in gesture, diagram, and word. In: Coventry KR, Tenbrink T, Bateman JA (eds) Spatial language and dialogue. Oxford University Press, Oxford, pp 119–131
56. Tversky B, Lee PU (1998) How space structures language. In: Freksa C, Habel C, Wender KF (eds) Spatial cognition: an interdisciplinary approach to representation and processing of spatial knowledge. Springer-Verlag, Berlin, pp 157–175
57. Tversky B, Lee PU (1999) Pictorial and verbal tools for conveying routes. In: Freksa C, Mark DM (eds) Spatial information theory: cognitive and computational foundations of geographic information science. Springer, Berlin, pp 51–64
58. Tversky B, Suwa M (2009) Thinking with sketches. In: Markman AB, Wood KL (eds) Tools for innovation. Oxford University Press, Oxford
59. Tversky B, Zacks J, Lee PU, Heiser J (2000) Lines, blobs, crosses, and arrows: Diagrammatic communication with schematic figures. In: Anderson M, Cheng P, Haarslev V (eds) Theory and application of diagrams. Springer, Berlin, pp 221–230
60. Tversky B, Agrawala M, Heiser J, Lee PU, Hanrahan P, Phan D, Stolte C, Daniel M-P (2007) Cognitive design principles for generating visualizations. In: Allen G (ed) Applied spatial cognition: from research to cognitive technology. Erlbaum, New Jersey, pp 53–73
61. Wexler M, Kosslyn S, Berthoz A (1998) Motor processes in mental rotation. Cognit 68:77–94
62. Wohlschlager A, Wohlschlager A (1998) Mental and manual rotation. J Exp Psychol Hum Percept Perform 24:397–314
63. Zacks J, Tversky B (1999) Bars and lines: a study of graphic communication. Mem Cognit 27:1073–1079

Part IV
Neuroscience—State-of-the-Art

Creative States: A Cognitive Neuroscience Approach to Understanding and Improving Creativity in Design

Evangelia G. Chrysikou

Introduction: Creativity as a Topic of Scientific Inquiry

In *The Principles of Psychology* William James argued, "Genius… means little more than the faculty of perceiving in an unhabitual way" (1890, *The Principles of Psychology*, vol. 2, p. 110). Since then, the topic of creative innovation—of breaking away from established ways of thinking to generate something new—has been of interest for generations of psychologists and educators. However, despite its high societal value, the scientific study of creativity has been far from the forefront of psychological research, largely due to the difficulties associated with the definition and criteria for creativity and the theoretical and methodological shortcomings of early attempts to study creative production in the laboratory [1].

Among the common misconceptions regarding creativity—that have likely undermined its status as a topic of scientific inquiry in psychology—is the view that creative products come about as the result of extraordinary abilities that are possessed only by a group of selected individuals [2]. In this paper, I will embrace an alternative position: that creative innovations are based on the individual's past experience and are the product of ordinary thought processes, such as memory, problem solving, imagining, and analogical reasoning; these cognitive tools are available to everyone and serve as the basis for all creative achievements [3–5]. Critically, these cognitive processes can be studied in the laboratory using established scientific methodologies, thus allowing one to determine cause-and-effect relationships regarding creative thought.

Another commonly held position about creativity is that novel products must be deemed valuable within their respective field to be considered creative [6]. However, for the scientific study of creativity it is important to differentiate between the novelty and the value of the creative product. Although novelty can be defined as the generation of an idea or product that a person has not produced before, value is defined relative to current social and historical circumstances, hence its definition can

E. G. Chrysikou (✉)
University of Kansas, Lawrence, KS, USA
e-mail: lilachrysikou@ku.edu

J. S. Gero (ed.), *Studying Visual and Spatial Reasoning for Design Creativity*,
DOI 10.1007/978-94-017-9297-4_14

change over time [2, 7]. This means that, if our operational definitions of creativity include the notion of value, then a product that is considered creative today, may not be considered creative 10 years from now, and vice versa (consider, for instance, that Egas Moniz received the Nobel Prize in Physiology and Medicine in 1949 for the invention of the prefrontal lobotomy, a procedure universally condemned today as inhumane; in contrast, the modernist painting *Nude Descending a Staircase Nº 2* by Marcel Duchamp is currently considered a modern art classic, yet was vastly rejected in its time). Accordingly, any conclusions drawn from experimental studies on creativity that confound novelty and value would not be generalizable to future creative acts. For these reasons, although acknowledging the pragmatic significance of the question of value for novel products, the research discussed here will examine creativity independently of the societal value of the creative product.

Finally, although creativity is frequently defined within the context of a specific domain (e.g., in the arts, music, literature, or science), in this paper I will define creativity more broadly as the goal-directed process of intentionally producing something novel that a given individual has not produced before. Notwithstanding that creative achievement is the result of a complex set of factors, including personal history, motivation, and personality characteristics, I will focus here on the cognitive processes underlying creative thought and their possible neural underpinnings.

Within this context, the research from cognitive psychology and cognitive neuroscience that will be presented in this paper has the potential to inform our understanding of creativity in many disciplines and could be applicable to investigators from a range of backgrounds and perspectives. Given the focus of this volume on visual and spatial creativity, however, I will discuss findings that pertain more closely to creative generation in design. Specifically, I will present research that (a) can promote our understanding of the cognitive processes implicated in design creativity and (b) may support educational efforts to foster creative production in design. These issues will be addressed in more detail in the following section.

Creativity in Design

Similar to the definition of creativity, the definition of design can take on different meanings for different individuals, depending on their focus on the design process or the design product. For the purposes of this paper, design refers to the cognitive process of generating and manipulating representations involved in solving a design problem within a given context and range of constraints [8–11]. In particular, during the design process, a designer (e.g., engineer, architect, product designer) uses his or her expert knowledge of (e.g., dimensions, appropriateness of style and materials) to generate and evaluate ideas toward the achievement of a goal (e.g., building a computer interface, designing an ecologically-friendly apartment building, generating a new travel mug), within a range of constraints (e.g., budget, space availability, client needs). As such, the design process involves a complex interplay between knowledge-driven or goal-driven (top-down) thinking and environmentally-driven or data-driven (bottom-up) thinking. Balancing *how* and *when* a designer moves from one end of this continuum to the other may predict his or her success in solving a design problem both creatively and efficiently [8].

Psychological studies of the creative process can illuminate the factors that determine a designer's flexibility in negotiating—on the one hand—their knowledge, past experience, and understanding of the design objectives, and—on the other hand—the need to come up with a design solution that is innovative and unique. Specifically with regards to the question of creativity in design problem solving, certain aspects of one's knowledge and experience may, actually, at times impede creative thought. At the same time, the generation of ideas that do not take into account design goals and constraints may undermine the designer's success in reaching a viable design solution.

Within this context, this paper will first review certain factors related to an individual's experience and knowledge organization that can impede creative idea generation and then discuss different techniques that have proven successful in addressing such constraints to creative thought. The following sections will further explore the hypothesis that the extent to which an individual uses their knowledge and experience during problem solving is associated with distinct patterns of brain activity that reveal a tradeoff between anterior prefrontal regions (typically involved in higher-order cognitive processing) and posterior, occipital-temporal regions (typically involved in visual object processing). Finally, based on evidence from cognitive neuroscience, the paper will offer the proposal that success in creative design is based on the designers' ability for prospective thinking and perspective taking—two cognitive processes that may allow good designers to predict the consequences of their design decisions and the audience's response to the newly-created products.

Design Fixation: Effects of Pictorial Examples on Creative Generation

Without doubt, one's knowledge and experience with certain kinds of problems or situations can support their attempts to solve a new problem that shares similar characteristics with the past. This phenomenon in problem solving is defined as *analogical transfer* and its positive effects have been well-documented in the literature [12, 13]. On the other hand, analogical transfer does not always promote successful solutions; in fact, reaching the solution to an earlier problem may have negative effects on current problem-solving attempts [14]. An example of such negative transfer is known as *functional fixedness* or *fixation* to a particular problem solving strategy that may not be useful in a current problem-solving situation [15, 16].

Functional fixedness is particularly pertinent to discussions of creative design due to its possible detrimental effects to the generation of creative design solutions. Of relevance to design education, in particular, is the phenomenon of fixation to pictorial examples during problem solving. Smith, Ward, and Schumacher [17], for instance, asked participants to imagine and create designs of different categories (e.g., animals to inhabit a foreign planet). In one of their conditions, participants were presented with pictorial examples prior to the design phase. Those subjects who were presented with the examples tended to reproduce in their sketches elements of the example designs, compared to subjects who were not shown such examples.

Fixation to pictorial examples has been specifically documented in the domain of engineering design [18–20]. Jansson and Smith [18] administered to engineering design students and professionals different design problems with the task to generate as many design solutions as possible. Although the total number of designs produced was similar, participants who had received example designs with the problems tended to conform to the elements of the example design significantly more so than participants who were not exposed to such examples. Critically, the effect did not diminish either when participants were given detailed descriptions emphasizing the negative characteristics of the example design, or when subjects were explicitly told to avoid replicating the examples. Interestingly, professional designers were not immune to this effect: they showed levels of fixation to examples that were comparable to engineering design students.

Purcell and Gero [19] examined in more detail the role of experience in a designer's susceptibility to fixation. They used a comprehensive coding methodology to examine the phenomenon across different designer disciplines and levels of expertise. Their findings replicated those of earlier research [18], though only for disciplines for which the example designs were characterized by an increased level of complexity (e.g., mechanical engineering). Based on these results, it is possible that the occurrence of fixation is determined by an interaction between a designer's discipline and the degree of complexity of the example design that may impose increased attentional demands on the designers.

The consequences of functional fixedness are critical for creative design. However, is design fixation observed exclusively in professional designers [21]? Or is it a broader cognitive phenomenon affecting design experts and novices alike? Moreover, are there instructional techniques that can be used effectively to eliminate design fixation?

In two experiments, Chrysikou and Weisberg [22] examined the occurrence of fixation to pictorial examples in participants who were naïve to design tasks. The participants were assigned to three conditions: (a) control (standard instructions), (b) fixation (inclusion of a problematic example, accompanied by description of its elements, including problematic elements), and (c) defixation (inclusion of a problematic example, accompanied by instructions to avoid using its problematic elements). Importantly, in contrast to prior work, participants were tested individually in a controlled laboratory setting. Participants saw multiple design problems during the session and were asked to generate as many ideas as they could for each, as well as draw sketches of their solutions and elaborate on their designs with brief comments. In addition, they were asked to read the task instructions aloud to ensure that they had reviewed and understood them in their entirety. Critically, in Experiment 1 participants were instructed to 'think aloud' during problem solving, so that a record of their thought processes could be obtained [23], whereas in Experiment 2 participants solved the problems silently.

Quantitative analyses focused on measures of design fixation as employed in previous work [19]. These included (a) measures of similarity (direct, reproductive, and analogical), (b) measures of reproduction of intentional flaws, and (c) measures of unintentional flaws. Moreover, to examine in detail the extent to which participants in each condition followed the examples, participants' verbal responses were transcribed and analyzed qualitatively by statement and for each problem separately. In particular, each statement was assigned to one of ten categories, adapted from prior research [24]: (1) using the problem instructions to implement a step; (2) using the example to implement a step; (3) using the problem instructions to repair an impasse; (4) using the example to repair an impasse; (5) using the problem instructions to check an action or a decision; (6) using the example to check an action or a decision; (7) following the example; (8) personal reference; (9) interaction with the experimenter; and (10) miscellaneous [22]. Two independent raters coded all responses on both quantitative and qualitative measures with high inter-rater agreement.

With respect to the quantitative measures, participants in the fixation condition in Experiment 1 produced significantly more elements of the example in their solutions and included more intentional and unintentional flaws in their designs relative to participants in the control condition. Importantly, however, participants in the defixation condition were able to thwart the deleterious effects of pictorial examples and their performance did not differ from the control condition. The qualitative results mirrored the pattern of the quantitative findings. Experiment 2, for which participants did not think aloud, replicated the quantitative results of Experiment 1; that is, the inclusion of the example design produced strong fixation effects; however, explicit instructions to avoid using the features presented in the examples also eliminated the fixation effect.

Overall, research on design fixation suggests that naïve participants and experts alike are susceptible to the effects of negative transfer in design problem solving. Strikingly, participants tend to fixate on pictorial examples and reproduce their elements, even in cases where the examples are explicitly described as problematic [18, 19, 22]. However, when participants attend to the defixation instructions in a controlled laboratory setting, they can successfully prevent the deleterious effects of the examples on the creative generation process.

Pictures and Words as Stimuli in Open-Ended Tasks

The findings discussed in the previous section have demonstrated that pictorial examples can be an obstacle to creativity in design problem solving. A question that arises from this research is whether design fixation occurs only under specific circumstances, within the context of specific types of problems. In other words, are certain types of problem solving tasks more susceptible to functional fixedness from pictorial stimuli than others?

Problem solving, in general, (and design problem solving in particular) refers to a situation in which the individual develops and implements plans with the intention of moving from a problem state to a goal state, within a range of constraints [2]. Some problems are well-defined or close-ended; for those, both the goal to be achieved and the path to be followed for the solution are obvious and the problem is perceived as having one correct answer (e.g., solving the equation $220 \times 3 = ?$). In contrast, other problems are ill-defined or open-ended; for those, the goal and the steps necessary for its completion are open to interpretation and the solution possibilities appear infinite. Consider, for instance Duncker's candle problem [15], a classic example from the problem solving literature: *Your goal is to attach a candle to a wall so that it can burn upright. You have available a candle, a book of matches and a box of tacks. How would you solve the problem?* The problem is vague and can have an infinite number of solutions (with the 'correct' one being to rethink or re-categorize the box of thumbtacks not as a container for the tacks but as a platform, tacking it to the wall and then placing the candle on the top).

Based on the findings from the design fixation literature as reviewed above, it is possible that the effects of pictorial stimuli are particularly strong for open-ended tasks (i.e., when multiple solutions are possible, e.g., designing a GPS system for the disabled), but are not equally present when the task is close-ended (i.e., when there is one correct answer, e.g., calculating the dimensions of a ceiling beam). In a recent study, Chrysikou and colleagues [25] explored this hypothesis by asking college students to respond verbally to one of three tasks that varied on this close- to open-ended dimension: (a) generate the typical use for a set of everyday objects; (b) generate a common alternative use for a different set of objects; and (c) generate an uncommon use for another set of objects. Critically, a third of the subjects were shown words as stimuli, a third of the subjects were shown pictures as stimuli, and the remaining third were shown both the word and the picture for each object.

Participants' verbal onset reaction times were recorded for quantitative analysis. In addition, their responses were transcribed and analyzed qualitatively with the use of a novel categorization system that categorized participants' answers for each object and task on a continuum from top-down-driven to bottom-up-driven responses. Specifically, top-down-driven responses were used to describe typical object functions (e.g., *chair:* to sit on) or functions that substituted the object for another tool based on shared abstract properties (i.e., properties not visible or available without prior knowledge of what the object is; e.g., hairdryer: to blow leaves). In contrast, bottom-up-driven responses were used to describe functions that substituted the object for another tool based on shared perceptual properties (i.e., properties visible or available without prior knowledge of the object's identity; e.g., tennis racket: to use as a snow shoe); they further described the generation of a new function for the object based on its bottom-up, perceptual properties (e.g., *chair:* to use as firewood).

According to the results, although there was no difference in reaction times by stimulus modality, participants who were exposed to the pictorial stimuli, produced significantly *less* bottom-up-driven and more top-down–driven responses than participants who were exposed to the word stimuli; however, this effect was obtained *only* when they performed the open-ended task, that is, when they generated uncommon alternative uses for the objects. Participants who were exposed to a combination of words and pictures did not differ in their responses with either of the other two conditions. These results demonstrate that pictorial stimuli can influence participants' performance in open-ended tasks significantly more so than in close-ended tasks. In particular, the presence of pictures increased the likelihood that participants, when generating uncommon uses for objects, produced uses that conformed to their knowledge of the object's canonical function.

Concepts, Categories, Goals, and Experience

Beyond the effects of stimulus modality as discussed above, our ability to categorize and re-categorize a tool depending on the context is an ability that we all share as goal-oriented beings. This ability is integral to us achieving goals and underlies our proficiency as toolmakers and innovators. What does this ability entail? What are the cognitive systems that allow for this flexible goal-oriented behavior? How can we facilitate optimal goal achievement in everyday problem solving tasks? This section of the paper will focus on the cognitive processes that allow us to employ specific aspects of what we know about an object (or our *semantic knowledge* for that object), as well as how we move beyond our typical interactions with it in order to achieve a given goal.

Categorization in Problem Solving

As mentioned earlier, every problem-solving situation, in which someone is using common objects to achieve a goal, can be described as the result of a continuous interaction between top-down (or knowledge-driven) and bottom-up (or stimulus-driven) processes. For example, if I am out of Styrofoam peanuts and I need to pack a gift, I could start from the goal and then think of ways to satisfy it by examining the properties of the objects around me that could work in that context (e.g., popcorn). Alternatively, I could start from the properties of the objects (e.g., popcorn) and then try to think of goals that they could serve. Although both processes are critical for problem solving, within the context of creative generation it is useful to examine whether one's reliance on preexisting knowledge may impede certain aspects of creative problem solving, and whether adopting a bottom-up (feature-driven) mode of thinking can enhance performance on creative generation tasks.

In particular, our long-term knowledge about the world (or our semantic knowledge) has been described in terms of a taxonomic organization, according to which knowledge is organized in distinct, category-specific domains (e.g., birds, mammals, vehicles) [26]. Here, I refer to semantic knowledge as a distributed knowledge system, according to which concepts are distributed across several interconnected domains based on concept attributes or properties (e.g., shape, size, color) that generally correspond to the brain regions originally involved in the acquisition of these properties (e.g., visual cortex for visual information, auditory cortex for acoustic information, and so forth) [27, 28].

Similarly, the term categorization is frequently used to refer to the organization of kinds in taxonomic categories [29]. Here, with the terms categorization or conceptualization I refer to the process of constructing a temporary working memory representation of a category that is derived from our long-term knowledge within a particular context (e.g., constructing a working memory representation of the concept 'fruit' within the context of purchasing at the grocery store fruit appropriate for a fruit salad). Accordingly, with the term concept I refer to the temporary construction in working memory that is used to represent a category on a particular occasion (e.g., in the example above, only certain aspects of one's knowledge of fruit, its appropriateness for a fruit salad, would be active in that context) [30–33].

If our knowledge about the world is organized in categories and concepts in a distributed fashion, how do people dynamically navigate this knowledge to interpret a problem situation and how do they use it to put together a successful strategy toward a goal? To address this question, I have argued that [34]: (1) When people attempt to achieve a goal they activate knowledge that is relevant to the achievement of that goal within that context and (2) the process of establishing relationships between one's knowledge and the information provided in the problem-solving situation involves numerous categorizations of the elements of a problem according to one's experiences. To clarify this position, consider the following example: a football is typically seen as "a ball with which you score a touchdown." However, seeing a football as "something that floats" becomes particularly relevant when one is drowning in a swimming pool. In contrast, the 'floatability' of the football would most likely not be a particularly salient component of our working representation of it in the middle of a football game [35]. Accordingly, being able to access the right kinds of information from one's knowledge of footballs within each context, may determine their success in using the object successfully to address each goal.

In practice, people can form taxonomic categories about items in the world by learning (and recreating) specific, idiosyncratically-interpreted exemplars from their personal experiences (e.g., fruit, clothes, furniture). Beyond these taxonomic categories, however, in the presence of an impromptu goal, people can construct goal-derived categories through the effortful, mostly top-down, and dynamic process of conceptual combination (e.g., things to sell at a garage sale, ways to make friends, things that can float). These goal-derived categories can be either well-established or ad hoc, depending on one's experience with the particular circumstances [35–37]. For example, if one is a frequent organizer of barefoot-bohemian-themed parties, the goal-derived category "activities for a barefoot-bohemian-themed party" will be well-established (i.e., it will be easy to instantiate that category with specific exemplars). Importantly, these instantiations can vary widely from individual to individual depending on the context and one's particular experiences (e.g., ways to make friends can mean different things for an incoming and an outgoing president) [35].

Categorization Training as a Way to Improve Creativity

With regards to creative problem solving, it is thus possible—based on what was discussed above—that success in goal-achievement depends on the individual's ability to break away from well-established categories and construct goal-derived categories, particularly those that are formed ad hoc to serve specific goals. Critically, individuals may differ in their ability to construct these categories depending on (a) whether the task is close-ended (e.g., frying an egg) or open-ended (e.g., constructing a survival kit for natural disasters) and (b) the individual's experiences and how these experiences match a given problem situation. Specifically, if in open-ended tasks the construction of goal-derived categories is critical but difficult to execute, it is likely that training participants to broaden their category boundaries may improve their performance in these tasks.

This hypothesis was examined in two experiments [38]. In the first study, participants were assigned to four conditions depending on the type of training they received: Participants in the Alternative Categories with Critical Items Task (ACT-C) condition generated as many as six alternative categories for 12 common objects. The training task included items critical for the solution to the problems used as dependent measures that were determined after norming (e.g., the tack box, in the Candle Problem). Participants in the Alternative Categories Task (ACT) condition received the ACT task, which was identical to the ACT-C with the difference that none of the objects included were relevant for the solution to the problems that followed (for example a newspaper has nothing to do with the solution to the Candle problem). Participants in the Embedded Figures Test (EFT) condition received the EFT, which was used as a control task to address whether any activity involving "flexible thinking" would work as training. In this task the subject has to identify a simple shape within a complex figure. Finally, participants in the Word Association Test (WA) control condition received as training a simple word association test. Immediately following training, all participants received seven open-ended problems like the Candle problem, all of which required creative problem solving involving everyday objects. Based on earlier research that has shown failure to transfer knowledge from one problem solving situation to another, unless explicitly told to do so [12, 14], for groups ACT, ACT-C, and EFT, participants received specific task instructions regarding the relevance of the training phase to the problem-solving phase.

As predicted, training with the ACT and ACT-C tasks significantly improved problem-solving performance: Participants in these two conditions showed significantly higher proportion of correct solutions relative to participants in the other two conditions, which did not differ from each other. Critically, this effect did not increase with specific training with the items that were crucial for the solution to the presented problems (i.e., in the ACT-C task). These findings were replicated in a second study, which was identical to the first with the exception that participants did not receive explicit instructions regarding the relevance of the pre-problem solving task to the problem-solving phase. In other words, the effect of the training was strong enough to overcome participants' likely tendency to avoid transferring strategies from one task to another without explicit instructions [38].

Overall, these experiments demonstrate that the way people organize and activate their knowledge about the world can determine their success during creative object use. Importantly, they suggest that training people to 'shake up' their categories through a brief conceptual exercise can expand their ability to move beyond well-established category boundaries and consider alternative interpretations of the problem elements that can facilitate creative solutions. Whether the benefits of this training generalize to other tasks is an empirical question that is currently under investigation; nevertheless, the effectiveness of the ACT task as training—which did not include items relevant to the dependent measures—would hold promise for the use of this procedure to enhance creative generation in a variety of design problem solving tasks.

Creative States: Prefrontal Cortex and Creativity

Why does asking people to think about concepts more broadly promote problem solving? Is this task—which forces people to challenge traditional category boundaries—associated with a particular neural state? Work in neuroscience has revealed the critical role of the frontal lobes in higher-order cognitive tasks, tasks in which one has to exercise a certain level of cognitive control over available information to achieve optimal performance. Such tasks involve, for example, holding in memory recently-presented information (e.g., the *n*-back task, in which one needs to remember a word or digit presented *n* trials back [39]), rule switching (e.g., the Wisconsin Card Sorting task, in which one has to monitor an implicitly changing rule to sort cards by color, quantity, or shape [40]), or resolving interference from unwanted information (e.g., the Stroop task, in which one is asked to name the ink color in which a color word is written when they don't match, e.g., saying 'red' for the word 'blue' written in red ink [41]). The prefrontal cortex (particularly the left ventrolateral prefrontal regions) has also been implicated in tasks that require participants to retrieve information from their knowledge about the world (e.g., retrieving a verb associated with an object or performing similarity judgments among items based on a particular property, like an object's color or function) [42, 43]. A distinctive feature of all such tasks is that they are close-ended, that is, they require one correct response the form of which is typically known to the participants. However, much of everyday problem solving and, particularly, design problem solving—as discussed above—is open-ended, that is, there is no obvious single response and the tasks seem to have multiple, equally likely solutions. Is the prefrontal cortex implicated similarly in open-ended and close-ended tasks?

Hypofrontality and Bottom-up Thought

Recent evidence from neuroscience studies involving both normal participants and patient populations would suggest that—in contrast to close-ended tasks—certain aspects of open-ended tasks might benefit from a tradeoff between regions involved in rule-based processing (i.e., prefrontal cortex) and regions involved in object processing, particularly processing of object attributes or features (i.e., visual cortex) [44, 45]. Activity in these distinct brain regions may be associated with different types of thought, namely knowledge-driven or goal-driven (top-down) thinking and environmentally-driven or data-driven (bottom-up) thinking. Specifically, the prefrontal cortex, predominantly in the left hemisphere, may support the construction of rules and regularities about the world from which one is *abstracting away* during development from low-level, 'raw' environmental data (e.g., learning that chairs are used for sitting regardless of their shape, size, or color) [46]. In contrast, focusing on low-level, 'raw' perceptual information in the environment (e.g., sounds, shapes, colors, materials) may involve activity in more posterior brain regions (i.e., occipitotemporal cortex). Importantly, depending on the close-ended or open-ended nature of the creative task, an individual may benefit from either top-down or bottom-up thinking for optimal performance, as supported by these distinct brain regions [45].

With regards to creative production, it can be argued that the generation of ideas within the context of an open-ended task (e.g., a creative design task) might involve a temporary distancing from knowledge-driven (top-down) thought—as guided by the prefrontal cortex—and a focus, instead, on data-driven (bottom-up) thought, as supported by posterior brain regions. In fact, evidence from neuroscience would suggest that lower activity in the prefrontal cortex (*hypofrontality*) as the result of disease or injury, may enhance one's ability for bottom-up cognitive processing. For example, patients with progressive aphasia, a neurodegenerative disease that targets selectively the patient's left frontal and temporal cortex, have been reported to exhibit increased levels of visual ability in spontaneous drawing or painting that they did not possess prior to their disease [48, 49]. Moreover, certain individuals with autism appear to outperform normal participants in reasoning tasks that require acute visual processing. This effect has been attributed to diminished lateral prefrontal cortex function in these individuals, in conjunction with increased brain activity in visual processing (i.e., occipital) regions [49]. Indeed, the suboptimal prefrontal functioning in autism may increase the availability of bottom-up, environmentally-driven information in these patients, which may allow some of them to become musical, mathematical, or artistic savants [50]. Finally, patients with focal strokes in the left prefrontal cortex have been shown to outperform normal participants in creative problem solving tasks that require breaking away from rule-based thinking [51].

The effects of hypofrontal cognitive states on enhanced perceptual processing have also been observed in normal subjects. Specifically, temporarily disrupting left prefrontal cortex activity using rapid transcranial magnetic stimulation (rTMS, a

Table 1 Neuroscience techniques for the study of creativity

Technique	Definition
fMRI	A non-invasive technique that measures changes in blood flow across the brain associated with neural activity during a given cognitive task
rTMS	A non-invasive procedure that can excite or inhibit neurons in a given brain region after the application of a strong electric current induced through a coil by rapidly changing magnetic fields
tDCS	A non-invasive procedure involving the application of small currents to the scalp for a few minutes through two surface electrodes that can modulate cortical excitability
EEG	A non-invasive technique that records the electrical activity across the scalp produced by neural activity in underlying brain regions

procedure that induces strong magnetic pulses to the scalp, thus altering the activity of underlying brain areas, see Table 1) can improve absolute pitch perception and number estimation in normal subjects [52, 53]. With regards to creative thinking, tasks that require broad conceptual associations have been linked to highly complex electroencephalogram (EEG; see Table 1) patterns across the entire brain, but also reduced activity in frontal brain areas [54]. Finally, a recent study that employed functional magnetic resonance imaging (fMRI; a procedure that allows researchers to acquire images of brain activity while participants perform various cognitive tasks, see Table 1) has shown hypofrontal neural profiles in professional musicians during jazz improvisation, but not during the reproduction of well-practiced musical sequences [55].

Overall, recent findings from neuroscience would suggest that reduced prefrontal cortex activity may facilitate certain aspects of perceptual processing and can shift the participant's focus from abstract, knowledge-based thinking to bottom-up, data-driven thinking.

Hypofrontal Cognitive States and Creative Generation

The findings discussed in the sections above would suggest the possibility of distinct neural states associated with top-down and bottom-up thinking. However, does the extent of prefrontal cortical involvement depend on the close-ended or open-ended nature of the task? Importantly, is performance in real-world, open-ended creative generation tasks associated with a distinct hypofrontal neurocognitive state?

A recent neuroimaging study attempted to explore this question and examined—more directly relative to previous work—the link between performance in open-ended, creative generation tasks and diminished prefrontal cortical functioning in normal subjects [56]. It was hypothesized that closed-ended tasks (i.e., having either one or a finite number of possible responses for which the search in conceptual space is deliberate) depend on the controlled retrieval of conceptual memory through the selection of one prepotent response that is facilitated by the left ventrolateral prefrontal cortex. Conversely, open-ended tasks (i.e., having an infinite number of possible responses, for which the search in conceptual space is

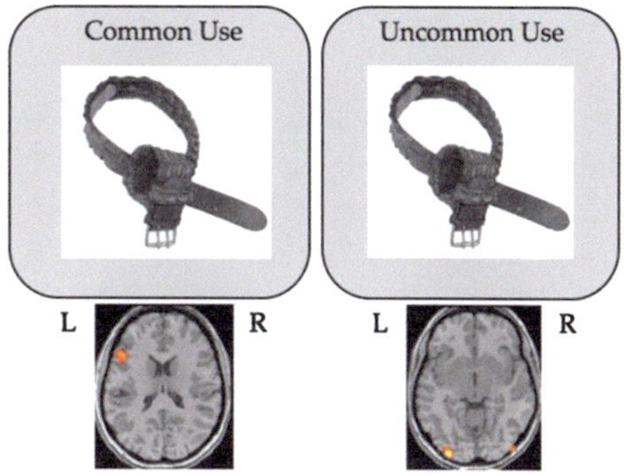

Fig. 1 Examples of stimuli and corresponding activations in the fMRI study [56]

non-deliberate) rely on the activation of posterior temporal-occipital regions specializing in object attributes or features within a distributed semantic network.

According to this prediction, we combined a close-ended task (i.e., common use generation, e.g., generating for *belt*: to keep one's pants up) and an open-ended task (i.e., uncommon use generation, e.g., generating for *belt*: to use as a tourniquet) in an fMRI paradigm to examine whether these tasks would lead to different types of response generation strategies. Participants were assigned to one of two conditions, depending on the task they had to perform (i.e., generate common or uncommon uses for everyday objects). They were shown grayscale images of the experimental stimuli and they were asked to generate aloud their responses while in an fMRI scanner. In line with our predictions, participants who generated the common use for everyday objects exhibited increased activity in left lateral prefrontal areas (see Fig. 1). In contrast, participants who generated uncommon uses for the objects did not show significant activation in prefrontal regions, but exhibited, instead, increased activation in posterior regions that are typically implicated in visual processing (left fusiform gyrus; see Fig. 1). Participants' responses in the uncommon use generation task were further transcribed and coded qualitatively on the continuum from top-down-driven to bottom-up-driven responses, as discussed earlier in this paper (see above, p. 8). These qualitative scores were then correlated with brain activation observed in each participant for an analysis of individual differences. According to the results of this analysis, the more participants' responses were categorized as perceptually based, the higher the activity in the middle occipital gyrus, a region involved in visual perception. This finding possibly reflects increased visual processing during this creative generation task.

In sum, this experiment has provided evidence for a tradeoff between regions involved in the controlled retrieval of conceptual information (i.e., prefrontal cortex) and those implicated in perceptual processing (i.e., posterior occipital regions). Specifically, these results demonstrate that in close-ended tasks, performance relies on the selection of appropriate information as facilitated by the prefrontal cortex; in contrast, in open-ended, creative generation tasks, in which the selection of one prepotent response would be counterproductive, diminished prefrontal cortical functioning, in conjunction with increased perceptual processing, optimizes performance.

Current and Future Directions

If hypofrontality states are associated with creative generation, is it possible to induce them artificially in normal subjects? As discussed above, rTMS has been used successfully to suppress transiently activity in the left prefrontal cortex, subsequently eliciting savant-like skills in healthy participants [52, 53]. A number of studies involving this procedure are currently underway to investigate the potential of this technique as a neuroenhancement tool for certain types of creative thought. However, rTMS is associated with high costs, difficulty of administration, and certain safety concerns. A different non-invasive procedure that addresses these problems is transcranial direct current stimulation (tDCS, see Table 1). tDCS introduces a brief electric current to the scalp and can modulate the excitability of neurons underlying the locus of stimulation. As such, the technique is currently used to inhibit cortical excitability in prefrontal and other cortical regions and explore the consequences of this modulation for cognitive function in a variety of creative thinking tasks. Although these techniques hold much promise for our understanding of creative cognition, among the aims of current and future research is to explore the magnitude, generalizability, and duration of the observed effects, as well as the effectiveness of these paradigms as interventions for the enhancement of creative thought.

Creativity as Prospective Thinking and Perspective Taking

In this paper I have approached creativity broadly as the process of generating something novel that results from the interplay between top-down, goal-driven thinking and bottom-up, data-driven thinking. Specifically, I have presented evidence that bottom-up thinking, as supported by a hypofrontal neural state, can promote the generation of creative ideas. Nevertheless, other aspects of the creative process can significantly benefit from top-down, knowledge-based thought. For example, evaluating design ideas for their appropriateness for a particular audience, or predicting the consequences of one's creative decisions before they are implemented, may determine the success or failure of a creative endeavor. Interestingly, recent work in neuroscience suggests that the human brain is constantly involved in this kind of prospective thought: a specific network of brain regions, including the dorsal prefrontal cortex, is continuously generating predictions about future events that are relevant for a given individual [57]. Critically, this is the same network of regions that is active when people are taking the perspective of another person within a specific context [58]. This ability for prospective thinking and perspective taking may be in the heart of the definition of creativity. Future research should focus on examining the involvement and importance of these brain circuits for our understanding of creative thought.

References

1. Sternberg RJ, Lubart TI (1999) The concept of creativity: prospects and paradigms. In: Sternberg RJ (ed) Handbook of creativity. Cambridge University, New York, pp 3–15
2. Weisberg RW (2006) Creativity: understanding innovation in problem solving, science, invention, and the arts. Wiley, Hoboken
3. Perkins DN (1981) The mind's best work. Harvard University, Cambridge
4. Simon HA (1988) Creativity and motivation. New Ideas Psychol 6:177–181
5. Weisberg RW (1993) Creativity: beyond the myth of genius. Freeman, New York
6. Csikszentmihalyi M (1996) Creativity: flow and the psychology of discovery and invention. Harper Collins, New York
7. Boden M (1990) The creative mind: myths and mechanisms. Weidenfeld and Nicolson, London
8. Lawson B (1997) How designers think: the design process demystified. (Rev. ed.). Architectural Press, England
9. Simon HA (1999) The sciences of the artificial, 3rd edn. MIT, Cambridge
10. Smith GJ, Gero JS (2004) The situated function-behaviour-structure framework. Des Stud 25:373–391
11. Gero J (1998) Towards a model of designing which includes its situatedness. In: Grabowski H, Rude S, Grein G (eds) Universal design theory. Shaker, Aachen, pp 47–56
12. Gick ML, Holyoak KJ (1980) Analogical problem solving. Cognitive Psychol 12:306–355
13. Holyoak KJ (1984) Analogical thinking and human intelligence. In: Sternberg RJ (ed) Advances in the psychology of human intelligence, vol. 2. Erlbaum, pp. 199–230
14. Gentner D (1983) Structure-mapping: a theoretical framework for analogy. Cognitive Sci 7:155–170
15. Duncker K (1945) On problem-solving. Psychological Monographs 58(5, Whole No.270)
16. Scheerer M (1963) Problem-solving. Sci Am 208:118–128
17. Smith SM, Ward TB, Schumacher JS (1993) Constraining effects of examples in a creative generation task. Mem Cognition 21:837–845
18. Jansson DG, Smith SM (1991) Design fixation. Des Stud 12:3–11
19. Purcell AT, Gero JS (1996) Design and other types of fixation. Des Stud 17:363–383
20. Purcell AT, Williams P, Gero JS, Colbron B (1993) Fixation effects: do they exist in design problem-solving? Environ Plan B: Plan Des 20:333–345
21. LeFevre JA, Dixon P (1986) Do written instructions need examples? Cognition Instruct 3:1–30
22. Chrysikou EG, Weisberg RW (2005) Following the wrong footsteps: fixation effects of pictorial examples in a design problem-solving task. J Exp Psychol: Learn Mem Cognition, 31:1134–1148.
23. Eriksson K, Simon HA (1993) Protocol analysis: verbal reports as data. MIT, Cambridge
24. VanLehn K (1998) Analogy events: how examples are used during problem-solving. Cog Sci 22:347–388
25. Chrysikou EG, Motyka K, Nigro C, Yang S, Thompson-Schill SL (2010) Functional fixedness for object use in open-ended tasks depends on stimulus modality. Submitted manuscript
26. Caramazza A, Mahon BZ (2003) The organization of conceptual knowledge: the evidence from category-specific semantic deficits. Trends Cog Sci 7:354–361
27. Tyler LK, Moss HE (2001) Towards a distributed account of conceptual knowledge. Trends Cog Sci 5:244–252
28. Thompson-Schill SL, Kan IP, Oliver RT (2006) Functional neuroimaging of semantic memory. In :R Cabeza R, A Kingston A (eEds)., Handbook of functional neuroimaging of cognition, 2nd edn. MIT, Cambridge, pp 149–190
29. Murphy GL (2002) The big book of concepts. MIT, Cambridge

30. Barsalou LW (1987) The instability of graded structure: implications for the nature of concepts. In: Neisser U (ed) Concepts and conceptual development: ecological and intellectual factors in conceptualization. Cambridge University, New York, pp 101–140
31. Barsalou LW (1989) Intraconcept similarity and its implications for interconcept similarity. In: Vosniadou S, Ortony A (eds) Similarity and analogical reasoning. Cambridge University, Cambridge, pp 76–121
32. Barsalou LW (1993) Challenging assumptions about concepts. Cognitive Dev 8:169–180
33. Simmons WK, Barsalou LW (2003) The similarity-in-topography principle: reconciling theories of perceptual deficits. Cognitive Neuropsychol 20:451–486
34. Chrysikou EG (2008) Problem solving as a process of goal-derived, ad hoc categorization. VDM, Saarbrücken
35. Barsalou LW (1982) Context-independent and context-dependent information in concepts. Mem Cognition 10:82–93
36. Barsalou LW (1983) Ad hoc categories. Mem Cognition 11:211–227
37. Barsalou LW (1991) Deriving categories to achieve goals. In: Bower G (ed) The psychology of learning and motivation: advances in research and theory, vol. 27. Academic, San Diego, pp 1–64
38. Chrysikou EG (2006) When shoes become hammers: goal-derived categorization training enhances problem solving performance. J Exp Psychol: Learn Mem Cognition 32:935–942
39. Owen AM, McMillan KM, Laird AR, Bullmore E (2005) N-back working memory paradigm: a meta-analysis of normative functional studies. Hum Brain Mapp 25:46–59
40. Buchsbaum BR, Greer S, Chang WL, Berman KF (2005) Meta-analysis of neuroimaging studies of the Wisconsin card-sorting task and component processes. Hum Brain Mapp 25:35–45
41. Banich MT, Milham MP, Jacobson BP, Webb A, Wszalek T, Cohen NJ, Kramer AF (2001) Attentional selection and the processing of task-irrelevant information: insights from fMRI examinations of the Stroop task. Cognitive Brain Res 134:459–470
42. Thompson-Schill SL, D'Esposito M, Aguirre GK, Farah MJ (1997) Role of left prefrontal cortex in retrieval of semantic knowledge: a re-evaluation. P Natl Acad Sci USA 94:14792–14797
43. Thompson-Schill SL, Bedny M, Goldberg RF (2005) The frontal lobes and the regulation of mental activity. Curr Opin Neurobiol 15:219–224
44. Dietrich A (2004) The cognitive neuroscience of creativity. Psychon B Rev 11:1011–1026
45. Thompson-Schill SL, Ramscar M, Chrysikou EG (2009) Cognition without control: when a little frontal lobe goes a long way. Curr Dir Psychol Sci 18:259–263
46. Wolford G, Miller MB, Gazzaniga M (2000) The left hemisphere's role in hypothesis formation. J Neurosci 20:1–4
47. Seeley WW, Matthews BR, Crawford RK, Gorno-Tempini ML, Foti D, Mackenzie IR, Miller BL (2008) Unravelling Boléro: progressive aphasia, transmodal creativity and the right posterior neocortex. Brain 131:39–49
48. Miller BL, Boone K, Cummings JL, Read SL, Mishkin F (2000) Functional correlates of musical and visual ability in frontotemporal dementia. Brit J Psychiat 176:458–463
49. Soulières I, Dawson M, Samson F et al (2009) Enhanced visual processing contributes to matrix reasoning in autism. Hum Brain Mapp 30:4082–4107
50. Snyder A (2009) Explaining and inducing savant skills: privileged access to lower level, less-processed information. Phil Trans R Soc B 364:1399–1405
51. Reverberi C, Toraldo A, D'Agostini S, Skrap M (2005) Better without (later) frontal cortex? Insight problems solved by frontal patients. Brain 128:2882–2890
52. Snyder A, Bahramali H, Hawker T, Mitchell DJ (2006) Savant-like numerosity skills revealed in normal people by magnetic pulses. Perception 35:837–845
53. Snyder A, Mulcahy E, Taylor JL, Mitchell DJ, Sachdev P, Gandevia SC (2003) Savant-like skills exposed in normal people by suppressing the left fronto-temporal lobe. J Integrative Neurosci 2:149–158

54. Mölle M, Marshall L, Wolf B, Fehm HL, Born J (1999) EEG complexity and performance measures of creative thinking. Psychophysiology 36:95–104
55. Limb CJ, Braun AR (2008) Neural substrates of spontaneous musical performance: an fMRI study of jazz improvisation. PLoS ONE 32:e1679
56. Chrysikou EG, Thompson-Schill SL (2011) Dissociable brain states linked to common and creative object use. Hum Brain Mapp 32:665–675
57. Bar M (2007) The proactive brain: using analogies and associations to generate predictions. Trends Cog Sci 11:280–289
58. Saxe R, Powell L (2006) It's the thought that counts: specific brain regions for one component of theory of mind. Psychol Sci 17:692–699

Spatial Transformations of Scene Stimuli: It's an Upright World

Amy L. Shelton and Jeffrey M. Zacks

At least two different types of mental spatial transformations can be used in spatial reasoning: object-based transformations—updating an object's spatial reference frame, and perspective transformations—updating the viewer's egocentric reference frame. Pictures of human bodies have been shown to flexibly engage these systems for different tasks, suggesting that the neural systems implementing these two transformations may be adapted for different spatial reasoning situations. In the present study, four experiments tested how pictures of immersive spaces—rooms—selectively engage different transformations. Response latency patterns suggested that the visual system quickly interprets pictures of scenes using two dissociable spatial transformations: object-based transformations, which re-orient the picture with respect to upright in the world, and perspective transformations, in which the viewer imagines themselves taking up a position within the depicted scene.

Author Notes
Amy L. Shelton, Department of Psychological & Brain Science, Johns Hopkins University. ashelton@jhu.edu
Jeffrey M. Zacks, Department of Psychology, Washington University at St. Louis. jzacks@artsci.wustl.edu
Authors had equal contributions to the work and are therefore listed in alphabetical order. We thank Jennifer Burrows, Dana Clark, Justin Lerner, Diana Pak, Margaret Sheridan, Jean Vettel, and Jason Yoon and for running the experiments.

A. L. Shelton (✉)
The Johns Hopkins University, Baltimore, USA
e-mail: ashelton@jhu.edu

J. M. Zacks
Washington University in St. Louis, St. Louis, USA

J. S. Gero (ed.), *Studying Visual and Spatial Reasoning for Design Creativity*,
DOI 10.1007/978-94-017-9297-4_15

Reasoning About Bodies in Space

Much of daily human activity involves reasoning about the changing relationships among one's body, other objects, and the world. Although several different types of mental spatial transformation may be possible, two distinct classes have been identified in the literature: object-based and perspective transformations [1]. Object-based transformations involve the mental rotation or manipulation of an object. This transformation is akin to a stationary observer "watching" a moving object in space. For example, when packing a car for a trip, one might imagine the different ways the suitcases can be turned to fit into the space. Perspective transformations are transformations of oneself in space. This is akin to an observer mentally transporting himself to a new perspective and "seeing" the world from this new view.[1] For example, when lecturing to a large class, the instructor typically faces the audience. When directing the audience to a particular portion of screen, the instructor may indicate a direction ("bottom right"). To specify the location correctly, the instructor may imagine what the screen looks like from the perspective of an audience member. Previous work has suggested that these two types of transformations are distinguishable both in terms of the behavioral profiles they produce [2] and the neural substrates that participate in them [3–6].

Behaviorally, object-based and perspective transformations have been distinguished by their temporal dynamics [2], and [7]. For example, when Wraga and colleagues asked participants to make spatial judgments about learned arrays, participants were either instructed to use array rotations (object-based transformations) or viewer rotations (perspective transformations). The results revealed that response latency increased as a function of orientation for the array rotations, whereas response latency for viewer rotations was flatter. As an alternative approach, Zacks and colleagues used different tasks to induce the transformations. Participants were asked to view images of bodies rotated in the picture plane. They were asked either to determine whether two bodies at different angular disparities had the same or different arm extended (same-different task) or to determine whether a single rotated body had its left or right arm extended (left-right task). The same-different task was hypothesized to induce object-based transformations because the judgment depends on the relationship between the two bodies irrespective of the observer. On the other hand, the left-right task was hypothesized to induce perspective transformations because left and right have clear meaning in the egocentric reference frame; by aligning oneself with the body stimulus, one can easily assess whether the extended hand is now on one's left or right. Response latency patterns supported this hypothesis. For same-different judgments response times increased monotonically with increasing stimulus orientation, replicating previous results for object-based transformations [8]. For left-right judgments response times were largely equivalent across different stimulus orientation, consistent with the pattern obtained when participants were explicitly instructed to perform perspective transformations with

[1] "Seeing" is not meant to restrict this to visual experience of space; the literature is agnostic as to whether the information available from an imagined perspective may be multimodal or amodal.

similar stimuli [9]. In a subsequent study, participants were given instructions that mismatched the hypothesized "natural" transformation for a given tasks; results revealed impaired performance and response latencies that resembled the other task [10]. These findings were further supported by participants' introspective reports.

Object-based transformations and perspective transformations have also been dissociated neurophysiologically. In numerous studies, object-based transformations, such as mental rotation, have been associated with the inferior parietal cortex, particularly in the right hemisphere [4, 11–22]. Although perspective transformations have received less attention in the literature, left posterior regions have been implicated in tasks that likely require perspective transformations (e.g., [23] and [24]). In a direct comparison of object-based and perspective transformations of bodies, [4], found that regions at the junction of the temporal, parietal, and occipital lobes (TPO) in the right hemisphere were selectively activated by object-based transformations [5]. Complementing this single dissociation, two studies using variants of the array and viewer rotation tasks found double dissociations between object-based and perspective transformations ([3] and [6]). In both studies, left TPO cortex was selectively activated by perspective transformations, and right parietal cortex was selectively activated by object-based transformations. (In [3], a number of other regions were selectively activated during object-based transformations as well.)

The existence of dissociations between object-based transformations and perspective transformations in their behavioral profiles and neural correlates has led to the suggestion that the brain has (at least) two systems for spatial transformations: one that supports object-based transformations and one that supports perspective transformations [1]. Zacks and Tversky, [10], proposed that the engagement of a particular system should depend not only on the task but also on the type of stimulus being manipulated. They contrasted bodies, which can move independently or serve as the source of viewpoint, with small inanimate objects, for which the independent movement or manipulation is far more common; rarely would one ask what the world looks like from an object's perspective. Consistent with the predictions, they found evidence that participants flexibly used perspective transformations or object-based transformations to make judgments about pictures of bodies, whereas participants depended heavily on object-based transformations when making judgments about manipulable inanimate objects.

Zacks and Tversky, [10], provided clear evidence for distinctions within the domain of discrete objects. However, spatial reasoning is not restricted to this class of stimuli and often entails making judgments in a multi-object environment (e.g., maneuvering a car in a parking lot). The present study was therefore designed to investigate spatial transformations of scene stimuli in the form of images of rooms. Unlike bodies or other objects, rooms are stationary, upright entities. They do not undergo movement, but can serve as the loci of potential perspectives. Thus, the tuning of the object-based transformation system would be expected to be relatively unresponsive during judgments about rooms, and the perspective transformation system would be expected to be responsive during any room judgments.

The four experiments described here were designed to test the hypothesis that rooms would selectively evoke the use of perspective transformations. We contrasted rooms to bodies, which have consistently shown both object-based and perspective transformations depending on the spatial reasoning task. Participants made same-different and left-right judgments about pictures of rooms and pictures of bodies. If the two systems for spatial transformations are readily available for either type of stimulus, one would expect that both stimuli would yield object-based performance for the same-different task and perspective performance for the left-right task [2]. Alternatively, if rooms selectively engage the perspective transformation system, one would expect that they would tend to produce flatter slopes than those observed with bodies for the same-different task.

Experiment 1

Experiment 1 was designed to compare performance on the same-different and left-right tasks separately for bodies and rooms. All participants performed both tasks with both types of stimuli, and the patterns of performance were examined in the context of object-based and perspective transformations. Bodies provided the control condition. Based on previous studies, we expected that participants would perform object-based transformations in order to solve the same-different task and perspective transformations in order to solve the left-right task. We therefore predicted increasing response times with increasing stimulus orientation in the same-different task, and flatter response latency profiles in the left-right task, replicating previous results.

For rooms, we hypothesized that participants would be less likely to perform object-based transformations, even in the same-different task. This led to the specific prediction that the relationship between stimulus orientation and response time during the same-different task would be weaker for rooms than for bodies. That is, we expected that the same-different task would show a flatter response latency curve for rooms than for bodies, and this curve should be similar to the response latency curve observed for the left-right task (for both rooms and bodies).

Method

Participants

Sixty-five participants (33 male) from the Stanford University community volunteered in return for experimental credits in Psychology courses. All participants reported normal or corrected-to normal vision and hearing.

Materials

Body stimuli were line drawing images of human bodies with one arm extended in two different poses. Images were created at 12 different picture plane rotations ranging from 0° (upright) to 330° in 30° increments. Room stimuli were created by first creating two different rooms in a desktop virtual reality program (Virtus Walk-though Pro, Virtus Corporation, Cary, NC). Virtual snapshots were then taken of each room with a plant placed on either the right or left side of a door in the center of the back wall of the room. Given that rooms are inherently scenes, the images had to be cropped. To prevent the image boundaries from providing a salient reference frame specifying orientation, the images were cropped in a circular window as if looking through a large porthole. Rooms were then rotated to create images from the same 12 orientations as for the bodies. For both types of stimuli, angular disparity in the left-right task corresponded to this angular disparity from upright. Half of the trials were right-handed (right arm extended or plant to the right), and half were left-handed. For the same-different tasks, the stimuli at the 12 different orientations were paired to create 12 different angular disparities ranging from 0° to 330°, in 30° increments (e.g., the 30° and 90° images might be paired to create a 60° test trial). Half the trials had the arm or plant on the same side, and half had it on different sides.

Procedure

All participants performed 112 trials in each of the four combinations of task (same/different and left/right) and stimulus (rooms and bodies), in a counterbalanced order on a Macintosh computer running PsyScope software [25]. Prior to testing, participants received instructions for each task in written form. In the same-different task they were told to press the left button for "same" and the right button for "different." In the left-right task they were told to press the left and right button for "left" and "right" responses, respectively. For bodies, left and right were defined by the arm of the figure, whereas in rooms participants had to determine where the plant would be upon entering the door. Participants were then given 10 practice trials that were identical to the actual trials just prior to each task. For each trial, a cue appeared ("Hit any button to go on"). A fixation cross appeared for stimuli for 1500 s followed by the test stimuli presented either in pairs (for same-different) or alone (for left-right) (see Fig. 1). If the response was correct, the computer indicated so with a pleasant tone and the trial ended. If the response was incorrect, the computer buzzed and the stimuli remained on the screen until the correct response was entered. Both the response latency (to the first response) and the accuracy were recorded.

Results

Three participants (2 male) were removed before analyses due to incomplete data or error rates exceeding 30 % in any task block or 15 % overall. For the remaining 62 participants, error rates were low in both judgment tasks (4.3 % for left-right, 5.5 % for same-different). The small task difference in errors was statistically significant, $F(1, 61)=4.68$, $p=0.03$, but did not interact with stimulus set, $F(1, 61)=0.002$,

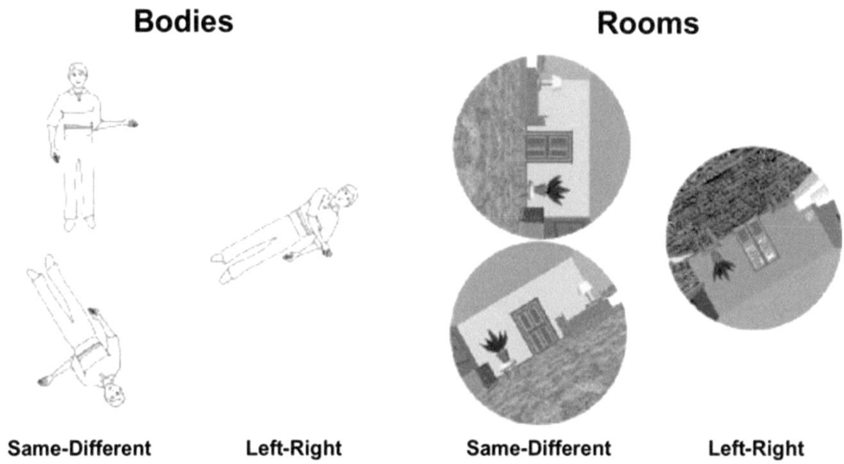

Fig. 1 Examples of the same-different and left-right tasks with pictures of bodies and rooms in Experiment 1. Room stimuli were presented in color during the actual experiment (Answers from left to right: same, right, different, left)

$p = 0.96$. The main effect of stimulus set on error rate did not approach statistical significance, $F(1, 61) = 0.0021$, $p = 0.96$.

All response time analyses were performed on correct responses only. Prior to analysis, outlying response times were trimmed by excluding observations 3 standard deviations from a participant's mean for a given combination of stimulus set and judgment task. This led to the elimination of 1.9 % of correct responses.

Two analyses were performed, following the approach of Zacks and Tversky, [10]. First, each participant's mean response times were calculated as a function of stimulus set, task, and orientation. These were then subjected to an analysis of variance (ANOVA) with stimulus type, task, and orientation as within-subject factors. As can be seen in Fig. 2, response times for same-different judgments about bodies increased substantially with increasing stimulus orientation, but response times for left-right judgments about bodies did not. For judgments about rooms, this difference was attenuated and both tasks showed smaller increases in response time with increasing stimulus orientation. This led to a statistically significant three-way interaction between stimulus orientation, task, and stimulus set, $F(6, 366) = 18.2$, $p < 0.001$. It also led to significant main effects of orientation, $F(6, 366) = 85.1$, $p < 0.001$, and of task, $F(1, 61) = 54.8$, $p < 0.001$, and to a significant two-way interaction between task and orientation, $F(6, 366) = 6.345$, $p < 0.001$. The two-way interaction between task and stimulus set approached but did not reach statistical significance, $F(1, 61) = 3.71$, $p = 0.059$. The main effect of stimulus set was not significant, $F(1, 61) = 1.15$, $p = 0.29$; nor was the interaction between stimulus set and orientation, $F(6, 366) = 0.73$, $p = 0.63$.

To more precisely characterize the relationship between stimulus orientation and response time we computed, for each participant, the Pearson correlation between orientation and response time for each combination of stimulus set and judgment

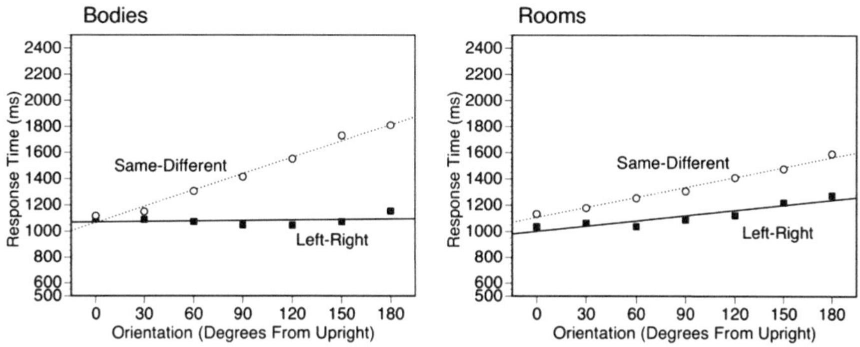

Fig. 2 Response time as a function of stimulus orientation for each combination of judgment (same-different or left-right) and stimulus set (bodies or rooms) in Experiment 1. Each point is the mean across participants of the mean within-participant trimmed response time. The lines are least-squared regression fits

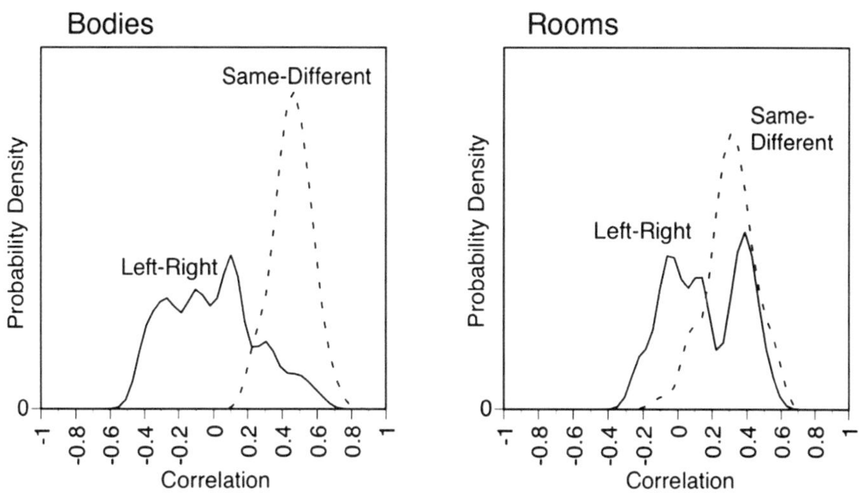

Fig. 3 Distributions of correlations between stimulus orientation and response time, as a function of the judgment made (same-different or left-right) and the stimulus set (bodies or rooms). Data are from Experiment 1. (For this figure and Figs. 5, 8 and 10, density functions were calculated by kernel estimation with a Gaussian kernel of bandwidth 0.05.)

task. The correlation gives a straightforward characterization of the strength of the linear relationship between stimulus orientation and response time. To the extent that response times increase with increasing orientation, this correlation will be positive (see [10]). The distribution of correlations for each condition is plotted in Fig. 3. As can be seen in the figure, for judgments about bodies, a clear task difference was observed: Correlations were robustly positive for same-different judgments, but centered on zero for left-right judgments. For judgments about rooms, correlations tended to be somewhat positive for both same-different and left-right judgments. This led to a significant main effect of task, $F(1, 61)=102.3$, $p<0.001$, and a significant interaction between stimulus set and task, $F(1, 61)=80.2$, $p<0.001$. The main effect of stimulus set was not significant, $F(1, 61)=0.11$, $p=0.74$. Follow-up t-tests revealed that the difference between the left-right and same-different tasks was significant for both the body stimulus set, $t(61)=13.2$, $p<0.001$, and for the room stimulus set, $t(61)=4.22$, $p<0.001$. Correlations were significantly positive for all conditions except for left-right judgments about bodies. [For that condition, $t(61)=-0.16$, $p=0.87$. For the other conditions, the smallest $t(61)$ was 5.7, $p<0.001$.]

To summarize, when making judgments about bodies, a strong difference was observed between same-different and left-right judgments: response time increased with increasing stimulus orientation for same-different judgments, but not for left-right judgments. However, for judgments about rooms, this task difference was attenuated; response time increased modestly for both left-right and same-different judgments. This led to a significant three-way interaction in the analysis of mean response times, and a significant two-way interaction in the analysis of correlations.

Discussion

The results of Experiment 1 were consistent with the hypothesis that room stimuli selectively engage perspective transformations: The relationship between stimulus orientation and response time in the same-different task was weaker for rooms than for bodies (flatter curve). However, there was an unanticipated pattern to these data. Although the stimulus orientation and response time relationship was attenuated when making same-different judgments about rooms, it was not fully orientation-independent; instead, response times increased significantly with orientation. Even more surprisingly, response times also increased significantly with orientation when making left-right judgments about rooms. Taken together, these results suggest that rooms differed from bodies in their engagement of object-based and perspective transformations, but they also raised questions about how rooms might show an attenuated increase in response time with stimulus orientation in both tasks.

One possibility is that participants performed perspective transformations for both the left-right and same-different tasks with rooms, but used trajectories that did not lead to the orientation-independent performance found for perspective transformations of front-facing bodies rotated in the picture plane [9]. The spatial framework of the room may have constrained participants' imagined perspective transformations, for example, if they imagined themselves deviating from the simplest path to avoid imagining themselves intersecting the objects near the door. On this interpretation, the data would provide support for the hypothesis that when participants thought of the room stimuli as immersive spaces, this produced a bias to solve the spatial reasoning problems using perspective transformations.

However, these data could also be explained by proposing that participants performed *object-based* transformations in both the left-right and same-different tasks with rooms. Perhaps presenting room stimuli as pictures induced participants to first resolve the discrepancy between the picture they were presented and the gravitational upright that they were experiencing by treating the picture of the room as an object unto itself and mentally rotating it to upright. Although experiences in which rooms rotate are presumably quite rare, experiences in which the reference frame of a room is misaligned with the gravitational upright are also atypical. In most cases where we see a room from an odd viewing angle, it is due to our own misorientation relative to the gravitational upright.

A third alternative is that the increased latency as a function of orientation reflects a natural tendency to upright a scene stimulus. Not only are actual rooms usually experienced in alignment with gravity, *pictures* of rooms are generally viewed such that the room depicted is aligned with the gravitational upright or the egocentric front of the viewer. (For example, paintings of rooms in museums generally are hung with the depicted floor and ceiling aligned with the actual floor and ceiling, and pictures of rooms in books are generally printed with the floor toward the bottom of the page and the ceiling toward the top.) As such, seeing a rotated scene stimulus may cause the participant to rapidly engage in an object-based transformation of the stimulus to reorient the depiction to upright, regardless of the task. By this explanation, there may be a bias to use perspective transformations for the spatial reasoning, but response latencies may be slowed down by the need to upright the image as well. In this sense, the object-based rotation of the depiction to upright is essentially interference.

In Experiment 2 we attempted to distinguish these three interpretations by directly instructing participants to perform perspective transformations with both body stimuli and room stimuli.

Experiment 2

To directly characterize the relationship between stimulus orientation and response time for perspective transformations with the room stimuli, we explicitly instructed participants to perform perspective transformations with those stimuli. Following the manipulations used by Parsons (1987), we asked participants either to perform the left-right task or to imagine a perspective transformation for both rooms and bodies. We predicted that participants performing the left-right task would show the same pattern of performance found in Experiment 1, with bodies showing a flat slope and rooms showing a slight increasing relationship. If participants who were asked to imagine performing perspective transformations showed the same pattern, this would support the hypothesis that the participants in Experiment 1 had tended to use perspective transformations when performing the left-right and same-different tasks with rooms. However, if participants who were asked to imagine performing perspective transformations showed orientation-independent performance for both bodies and rooms, this would suggest the participants in Experiment 1 had tended to use *object-based* transformations when performing the left-right and same-different tasks with rooms.

Method

Participants

Thirty-two participants (16 male) from the undergraduate population at Washington University volunteered in return for \$ 10 or partial fulfillment of a course requirement.

Materials

The materials were the same room and body stimuli used in Experiment 1.

Procedure

Participants were randomly divided into two groups and asked to perform two different tasks. In the *left-right* group, participants performed the left-right task described in Experiment 1. In the *imagine* group, participants were asked simply to "imagine [themselves] standing in the door of the room," and "form a vivid mental picture of [themselves] lined up with the door as shown on the screen." They were instructed to press a button when they had formed the image. Participants in each group performed 112 trials with each type of stimuli. The room and body blocks were counterbalanced across participants. Stimuli were presented in the same manner as in Experiment 1, except that there was no correct or incorrect response in the imagine task, and so no feedback was provided.

Results and Discussion

For the group that performed the left-right task, error rates were comparable to those in Experiment 1. They were low (4.6 % for bodies, 3.9 % for rooms) and did not differ significantly across stimulus sets, $t(15)=0.91$, $p=0.38$. Response time data from error trials were excluded, and the response time data were trimmed as described in Experiment 1, which resulted in the elimination of 1.5 % of correct responses.

The response time data were analyzed using the same approach as for Experiment 1. First, each participant's mean response times were calculated as a function of group, stimulus set, and orientation, and these were submitted to an ANOVA with group as a between-participants factor and stimulus set and orientation as repeated measures. As Fig. 4 shows, response time was relatively independent of orientation for both types of judgment about bodies, but increased somewhat with increasing orientation for both types of judgment about rooms. This led to a significant main effect of orientation, $F(6, 180)=3.12$, $p=0.006$, and a significant orientation-by-stimulus set interaction, $F(6, 180)=8.25, p<0.001$. Performance of the imagine task

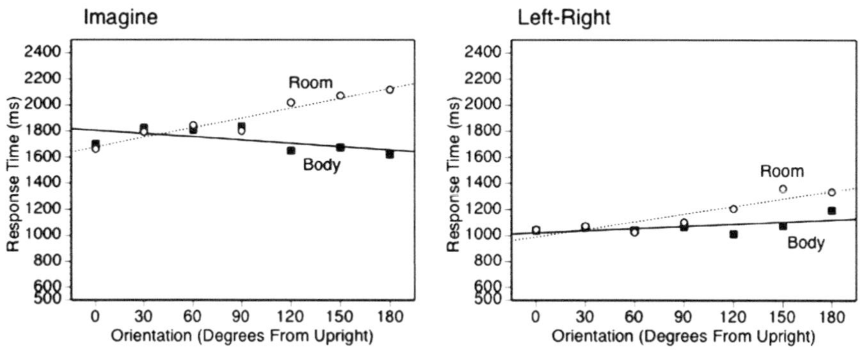

Fig. 4 Response time as a function of task (imagining oneself in the position indicated by the picture, or making a left-right judgment about the picture) and stimulus set (bodies or rooms) in Experiment 2. Each point is the mean across participants of the mean within-participant trimmed response time. The lines are least-squared regression fits

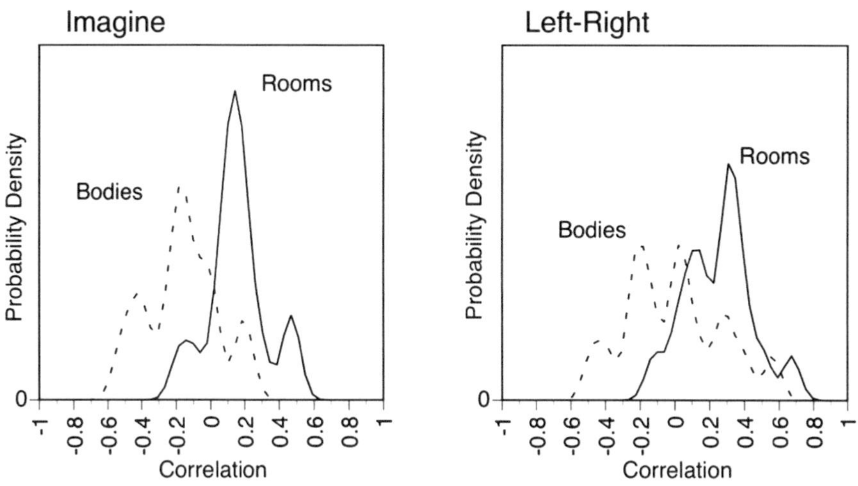

Fig. 5 Distributions of correlations between stimulus orientation and response time, as a function of task (imagining oneself in the position indicated by the picture, or making a left-right judgment about the picture) and stimulus set (bodies or rooms) in Experiment 2

was overall slower than performance of the left-right task, leading to a significant main effect of group, $F(1, 30)=4.93$, $p=0.034$. None of the other main effects or interactions approached statistical significance (largest $F=1.70$). To follow up the significant two-way interaction, we conducted separate ANOVAs for each of the two groups. These showed that the orientation-by-stimulus set interaction was significant for both the left-right group [$F(6, 90)=4.23$, $p<0.001$] and the imagine group [$F(6, 90)=5.15$, $p<0.001$]. Separate ANOVAs for each of the four combinations of group and stimulus set showed that the effect of orientation was statistically significant for both the left-right and imagine tasks with rooms [left-right: $F(6,$

90) = 6.79, $p<0.001$; imagine: $F(6, 90)=3.30$, $p=0.006$]. There was no significant effect or orientation for either task when performed with pictures of bodies [left-right: $F(6, 90)=0.99$, $p=0.44$; imagine: $F(6, 90)=1.74$, $p=0.12$]

To further characterize the relationship between orientation and response time across the experimental conditions, we calculated the correlation between orientation and response time for each participant, for each of the two stimulus sets. As can be seen in Fig. 5, correlations for both groups were higher when making judgments about pictures of rooms than when making judgments about pictures of bodies. This led to a significant main effect of stimulus set, $F(1, 30)=41.5$, $p<0.001$. Correlations also were slightly higher for the group that performed the left-right task, leading to a marginally significant main effect of group, $F(1, 30)=3.86$, $p=0.06$. The group-by-stimulus set interaction did not approach statistical significance, indicating that the two groups showed similar stimulus set effects, $F(1, 30)=0.46$, $p=0.51$. T-tests confirmed that for both groups, the difference in correlations between the two stimulus sets was significant [left-right: $t(15)=4.14$, $p<0.001$; imagine: $t(15)=4.96$, $p<0.001$]. Correlations between orientation and response time for judgments about room pictures were significantly positive, $t(31)=6.03$, $p<0.001$. For judgments about bodies, the correlations were slightly negative, and this difference approached statistical significance, $t(31)=-1.82$, $p=0.08$.

In sum, response time patterns when participants were explicitly asked to imagine themselves in a particular position were similar to response time patterns when participants were asked to make left-right judgments about the same position. For judgments about bodies, response time was essentially independent of stimulus orientation both when participants made left-right judgments, and when they were explicitly instructed to imagine themselves in the position of the body, replicating previous findings [9]. For judgments about rooms, response times increased with increasing stimulus orientation, both for left-right judgments and for imagined movements. These nearly identical patterns replicate those observed for the left-right task in Experiment 1, ruling out the possibility that participants were using strictly object-based transformations on the left-right task.

However, these data do not explain why these perspective transformations for rooms should be more sensitive to orientation than perspective transformations of bodies. More specifically, why should the time to imagine one's self in the door of a room should differ from the time to imagine one's self in that same position when the to-be-assumed position is cued by a picture of a body standing alone? The interference explanation introduced previously may account for this oddity. That is, when shown a depiction of a room in an atypical orientation, participants may perform an object-based transformation to upright the stimulus *in addition* to the transformations that are required for appropriately completing that task. By this explanation, the representation of the space depicted by the picture evokes a tendency to perform a perspective transformation, but the representation of the picture as a picture evokes a tendency to upright the picture using an object-based transformation. If this is correct, then the surface properties of the room pictures should be necessary and sufficient to evoke the object-based uprighting transformation.

Fig. 6 Example of the combined room/body pictures used in Experiment 3. Stimuli were presented in color in the actual experiment

Experiment 3 provided a rigorous test of the interference hypothesis using exactly the same stimuli to depict rooms and bodies. In this experiment, participants made left-right or same-different judgments about pictures that included both a body and a room, but were instructed to attend either to the spatial reference frame of the body, or of the room.

Experiment 3

Experiment 3 replicated Experiment 1, except that the stimuli were identical in the rooms and bodies conditions. We created new stimuli that included a body standing in a room (Fig. 6), and then manipulated the instructions to direct attention either to the room or to the body. These instructions did not tell the participant what type of transformation to use, but rather indicated what aspect of the stimulus (the room or the body) was relevant to the task.

By holding the physical stimuli constant, Experiment 3 allowed us to directly test what gave rise to the differences observed for rooms versus bodies. First, if the difference in response latency patterns between rooms and bodies on the same-different task resulted from the preferential engagement of perspective transformations when reasoning about rooms compared to bodies, then the response latencies should again show a more pronounced monotonic relationship to orientation for bodies than for rooms. Second, holding the stimulus constant allowed us to test how the stimulus differences may have affected the patterns of performance, particularly on the left-right task. In Experiment 2, when participants were asked to perform perspective transformations with rooms, small but significant increases in response time with increasing stimulus orientation were observed. In Experiments 1 and 2, response time increased slightly but significantly with increasing orientation for left-right judgments. This result differed from the pattern observed for left-right judgments about bodies in the same spatial configuration, in Experiment 1 and previous research [2] and [9]. We hypothesized that room stimuli might invoke some automatic transformation to upright, irrespective of the reference frame for making the judgment. Based on this hypothesis, we predicted that in the left-right task in Experiment 3, we would observe small but significant increases in response time with increasing stimulus orientation for *both* the body and room conditions.

Method

Participants

Sixty-four participants (32 male) from the Johns Hopkins Community volunteered in return for extra credit in Psychology and Cognitive Science courses or for monetary compensation.

Materials

Using Poser 3.0 software (Curious Labs, Santa Cruz, CA), rendered images of rooms (2 different rooms) and bodies (2 different poses) were created. In the images, a lamp was placed either to the left or right of the doorway and the body had either the left or right arm extended (see Fig. 6). The two rooms and two poses were combined such that room, pose, left or right lamp, and left or right arm were completely counterbalanced. As in the previous experiments, the images were cropped in a circular aperture, and images were taken at 12 different orientations ranging from 0° (upright) to 330° in 30° increments. These images were combined to create the different angular disparities for the same-different task.

Procedures

All participants performed the same-different and left-right tasks with both rooms and bodies, completing 112 trials of each combination. The trials were blocked hierarchically, first by attentional instruction and then by task. Participants were assigned to groups according to the complete counterbalancing of instruction and task within instruction. For the attend-rooms instructions, participants were asked to determine whether the lamp was on the same side of the door in two images in the same-different task and asked to determine whether the lamp was on the right or left of the door when entering the room in the left-right task. For the attend-bodies instructions, participants were asked to determine whether the two figures had the same arm extended in the same-different task and asked to determine which of the figure's arm was extended in the left-right task. The correspondence between the

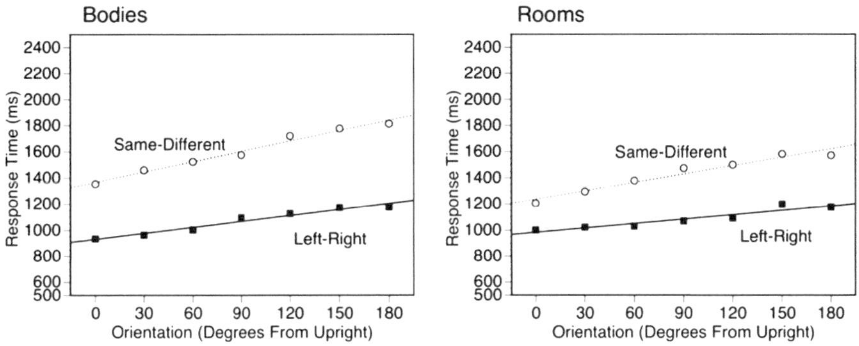

Fig. 7 Response time as a function of stimulus orientation for each combination of judgment (same-different or left-right) and object about which the judgment was made (bodies or rooms) in Experiment 3. Each point is the mean across participants of the mean within-participant trimmed response time. The lines are least-squared regression fits

location of the lamp and the extended arm was counterbalanced, such that attending to the wrong stimulus would produce chance performance. Trial procedures were identical to those used in Experiment 1.

Results and Discussion

Error rates were low (4.9 %) and did not differ significantly across conditions. [There was a marginally significant task-by-stimulus set interaction, such that error rates were slightly lower in the left-right task with bodies than the other three conditions, but this did not reach statistical significance, $F(1.64)=3.63, p=0.06$. Neither main effect was significant: For the effect of task, $F(1, 63)=1.04, p=0.31$; for the effect of stimulus set, $F(1, 63)=1.96, p=0.17$.]

Response time data were trimmed and analyzed as described for Experiment 1. First, mean response times were calculated for each participant for each combination of task, instructions, and orientation, and these mean response times were submitted to a repeated measures ANOVA. As can be seen in Fig. 7, when participants attended to the bodies there was a large difference between the same-different and left-right tasks, such that response times increased more with increasing orientation during the same-different task. When participants attended to the space of the rooms, this difference was attenuated. This pattern led to a three-way interaction between task, instructions, and orientation, $F(6, 378)=3.46$, $p=0.002$, and replicated the pattern observed in Experiment 1. However, response times increased with increasing orientation for all four conditions, including a small but significant increase for left-right judgments when attending to the body [overall $F(6, 378)=71.6, p<0.001$; smallest individual-condition $F(6, 378)=10.2$, $p<0.001$]. Overall, responses were slower in the same-different task, $F(1, 63)=85.8$, $p<0.001$, and slower when attending to the bodies than when attending to the space of the rooms, $F(1, 63)=7.17$,

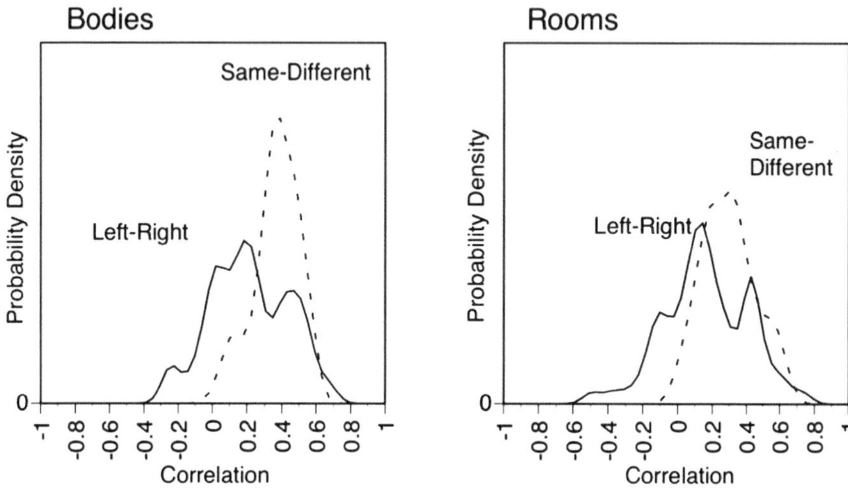

Fig. 8 Distributions of correlations between stimulus orientation and response time, as a function of the judgment made (same-different or left-right) and the object of the judgment (bodies or rooms) in Experiment 3

$p=0.009$. All three two-way interactions were also significant, smallest $F=9.68$, $p<0.001$.

Analyses of the correlations between stimulus orientation and response time largely converged with the ANOVAs on response time. Correlations were significantly higher for same-different judgments than for left-right judgments, $F(1, 63)=26.5$, $p<0.001$, and were higher when participants attended the bodies than when they attended to the rooms, $F(1, 63)=16.7$, $p<0.001$ (Fig. 8). For all four conditions, the correlations between stimulus orientation and response time were significantly positive, smallest $t(63)=5.46$, $p<0.001$. However, the correlation analyses failed to provide additional evidence that the relationship between stimulus orientation and response time depended on the interaction of task and instructions; this was not statistically significant, $F(1, 63)=0.69$, $p=0.41$

In short, the results replicated the main finding of Experiment 1: The relationship between stimulus orientation and response time depended both on the judgment participants were asked to make, and on the target of that judgment.

The same-different task revealed the same pattern of stronger orientation dependence for bodies than for rooms, arguing against the possibility that this difference was due to stimulus differences alone in Experiment 1. This predicted difference supports the hypothesis that participants used more perspective transformations and fewer object-based transformations when making same-different judgments about rooms compared to bodies. The same-different judgments for rooms as well as the left-right judgments for both stimuli have patterns nearly identical to that observed in Experiment 2, when participants were directly instructed to imagine making a perspective transformation and cued with a picture of a room. This pattern further supports the preferential use of perspective transformations for bodies in the left-right task and rooms more generally.

The weak but significant increase in response latency as a function of stimulus orientation in both versions of the left-right task suggest an influence of the room stimulus, irrespective of the focus of the transformation. This finding is consistent with the hypothesis that pictures of rooms at atypical orientations tend to evoke object-based transformations to mentally upright the pictures, in addition to perspective transformations that may be performed to accomplish the left-right judgment.

The strong influence of the room in the bodies condition supports the claim that the uprighting is occurring in a task-irrelevant manner. However, it is notable that the rotation of these stimuli was locked such that the body and room rotated together. If pictures of rooms at atypical orientations evoke object-based transformations to upright them, and if people also tend to perform perspective transformations to make left-right judgments about a potential viewpoint from a body within a room, then manipulations of the room's orientation and the body's orientation should have separable effects: in this paradigm, response times should increase with increasing rotation of the room, but not with increasing rotation of the body. Experiment 4 provided a stronger test of the task-independent uprighting account by asking participants to make left-right judgments about bodies only and rotated the room or the body independently.

Experiment 4

If effect of orientation on response time with room stimuli across both tasks reflects a task-irrelevant tendency to upright a room stimulus, then this effect should occur even if the room rotation is independent of the body that is being judged. To test this, participants in Experiment 4 were asked to make left-right judgments about bodies only while we varied the relationship between the rooms and bodies separately. In the *body-rotate* condition, the room was maintained in the upright position in the background and the body was rotated. In the *room-rotate* condition, the body remained in the upright position and the room was rotated in the background (Fig. 9).

If the task-independent uprighting hypothesis is correct, then performance should be orientation independent when the room is upright and the body is rotating, just as in the conditions where the body is presented alone, whereas performance should be orientation dependent when the room is rotating even though the body about which the judgment is made remains in the upright position. By contrast, if the participants can ignore the task-irrelevant room rotation, then both conditions should produce patterns identical to the bodies alone in Experiment 1. Finally, our stimuli could introduce a third type of discrepancies by having the bodies and rooms in inconsistent orientations. If the "uprighting" tendency is sensitive to any type of incongruence, then both conditions might show orientation dependence as the participant attempts to reconcile the angular disparity between the room and body.

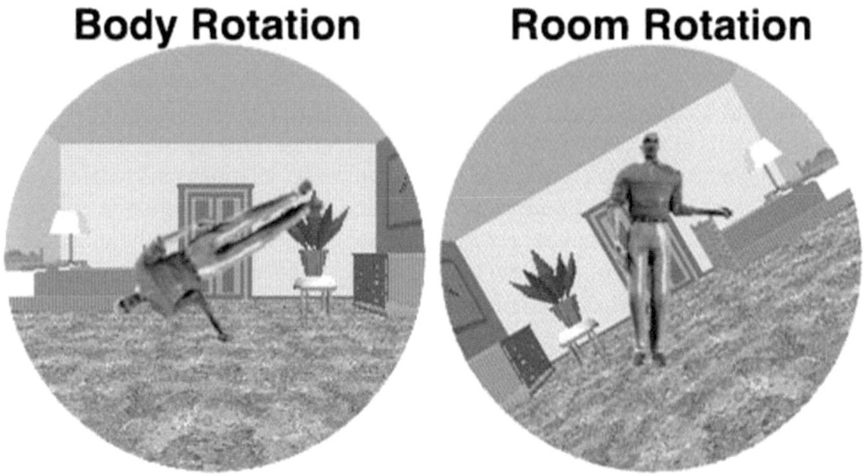

Fig. 9 Examples of the stimuli used in Experiment 4 showing a body rotating against a stable upright room and a stable upright body against a rotating room

Method

Participants

Twenty-four participants (12 male) from the Johns Hopkins Community volunteered in return for monetary compensation.

Materials

Using the same room images from Experiment 1 and bodies created as in Experiment 3, images were created that had the bodies in front of the doors of rooms as in Fig. 6. We used four base images that counterbalanced whether the extended hand of the body was on the same side as the plant in the room, even though participants were never asked about the plant (or any other feature of the room). From these four base images, two sets of stimuli were created. For the body rotation conditions, the room remained upright and the body in front of the door was rotated in the 12 different orientations ranging from 0° (upright) to 330° in 30° increments. For the room rotation condition, the body remained upright and the room in the background was rotated in the same 12 orientations. The 0° images for the two conditions were identical, so trials were randomly designated as belonging to one condition or the other to maintain independence of the two conditions.

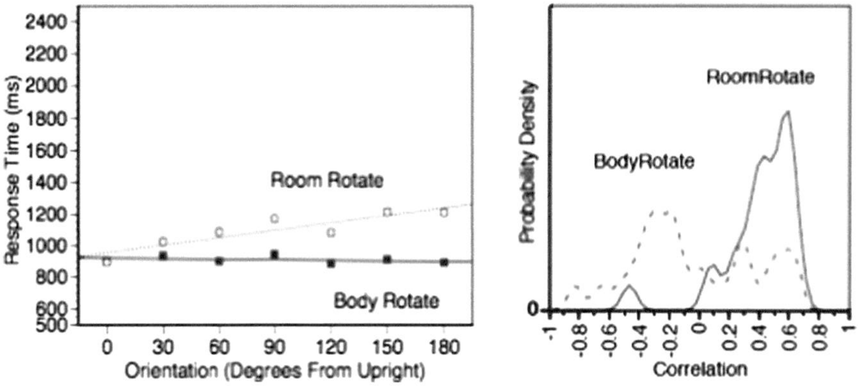

Fig. 10 Data from Experiment 4: Top panel shows response latency as a function of the orientation of either the body (closed squares) or the room (open circles) for the left-right task. Bottom panel shows the distribution of correlations between stimulus orientation and response time as function of which part of the stimulus was rotating (body or room)

Procedures

All participants performed left-right task on the bodies only using both sets of stimuli. Stimuli from the body rotation and room rotation conditions were presented in random order, and conditions were not explicitly revealed to the participants. Trial procedures were identical to the left-right task used in Experiment 1.

Results and Discussion

Error rates were low—3.5 and 2.4 % for the body and room rotation, respectively. Response time data were trimmed and analyzed as described for Experiment 1. First, mean response times were calculated for each participant for each combination of condition and orientation, and these mean response times were submitted to a repeated measures ANOVA. As shown in the left panel of Fig. 10, there was a pronounced condition-by-orientation interaction, $F(6, 138)=6.87$, $p<0.001$, with response latency showing a stronger linear relationship with the room rotations than with the body rotations, $F(1, 23)=21.9$, $p<0.001$. Overall, responses were slower in the room rotation condition, $F(1, 23)=34.7$, $p<0.001$, and showed orientation dependence, $F(6, 138)=4.91$, $p=0.003$ [Linear contrast, $F(1, 138)=12.2$, $p=0.002$]. However, these effects were likely due to the interaction. The correlation between orientation and response time was greater for the room rotations than the body rotations, $t(23)=4.30$, $p<0.001$. Moreover, the average correlations were -0.04 and 0.41 for the body and room rotations, respectively, supporting the observation that the room rotations showed a substantial influence of orientation (Fig. 10, right panel).

These results support the hypothesis that rotated scene stimuli—even when they are task irrelevant—invoke some degree of automatic transformation to upright the world. In the body rotation condition, participants appeared to use perspective transformations; neither the rotation of the body relative to upright nor the discrepancy between the irrelevant room stimulus and the body affected response times. However, in the room rotation condition, the to-be-judged body stimulus was always upright with respect to the participant (and the computer screen, the testing room, etc.), but response times were affected by the rotation of the irrelevant room stimulus in the background, supporting the hypothesis that some task-irrelevant transformation is occurring in response to the presence of a rotated scene stimulus.

General Discussion

The four experiments reported here tested the degree to which scene stimuli (rooms) preferentially engaged perspective transformations more than object-based transformations. Previous research [10] has suggested that people tend to perform object-based transformations when making judgments about pictures of small, manipulable objects. The present results argue that people tend to perform perspective transformations when making judgments about pictures of scenes. This pattern is consistent with people's everyday experience of objects and places: Objects often move around us or are moved by us, and it is important to predict the consequences of those movements. Places, however, are generally stable. For places it is important to predict the consequences of occupying one location or another within the space. Bodies occupy a unique intermediate role: We experience them both as objects that can move around, when we watch other people, and as cues to potential locations of perspective, when we ourselves move around in the world. Consistent with this dual role, in these experiments and in previous studies, [2] and [10], when cued with a body, participants appeared to be able to flexibly perform either an object-based transformation or a perspective transformation, depending on the spatial judgment that needed to be made.

Experiments 1 and 3 provided evidence that spatial judgment response times depend on both the spatial judgment one is making and the thing about which that judgment is made. For the same-different task, there was a relationship between stimulus orientation and response, consistent with the performance of object-based transformations. However, this relationship was stronger for bodies than for rooms, consistent with the hypothesis that participants would be less inclined to use object-based transformations when reasoning about the room stimuli. A substantially weaker relationship was observed for both types of stimuli in the left-right task, supporting the use of perspective transformations, as expected.

In addition to the robust difference between rooms and bodies, there was a small but consistent effect of orientation on response latency for room stimuli in both tasks such that the response latency patterns for room stimuli were neither strongly linear (as expected for object-based transformations) nor orientation-independent.

Instead, for both the same-different and left-right tasks, we observed an attenuated trend for increased response latency as a function of angular disparity.

When presented with a picture of a room at an orientation that conflicts with other salient reference frames, participants may initially perform an object-based transformation of the picture to bring it into alignment with those other reference frames, independent of the spatial judgment task. Unlike pictures of bodies, pictures of rooms include salient straight lines and 90° intersections, establishing the planes of the walls. These features are strong cues to the reference frame of the picture. When room pictures are rotated, that reference frame conflicts with the reference frames defined by the participant's eye position, the computer screen, the room in which the experiment takes place, and gravity. The fact that response times for pictures of rooms increased less with orientation than response times for pictures of bodies, and were less affected by task manipulations than were response times to body pictures, argues for the view that participants tended to solve problems involving pictures of rooms by performing a perspective transformation to place themselves in the position depicted by the room. The results of Experiment 4 suggest that this uprighting need not be relevant and may not be requisite for the actual judgment but is an interference occurring in a more automatic fashion any time a rotated scene is presented. Recent studies provide additional evidence for the uprighting hypothesis by identifying the reference frame(s) used to define upright for scenes [26].

Together, these data provide clear evidence that performance in spatial reasoning tasks depends both on the type of spatial judgment required and on the stimulus about which the judgment is made. In particular, participants showed evidence of a tendency to use perspective transformations when reasoning about room stimuli, even for same-different judgments, which strongly evoke object-based transformations when made about pictures of bodies [2] and [10]. This interaction of task and stimulus provides compelling support for the view that multiple spatial transformation systems are tuned to be responsive to the requirements of different spatial reasoning situations. The adaptive deployment of these computational tools may form building blocks for complex skills such as navigation, long-term spatial memory, and abstract reasoning.

References

1. Zacks JM, Michelon P (2005) Transformations of visuospatial images. Behav Cogn Neurosci Rev 4:96–118
2. Zacks J, Mires J, Tversky B, Hazeltine E (2002) Mental spatial transformations of objects and perspectives. Spat Cogn Comput 2:315–332
3. Wraga MJ, Shephard JM, Church JA, Inati S, Kosslyn SM (2005) Imagined rotations of self versus objects: an fMRI study. Neuropsychologia 43:1351–1361
4. Zacks J, Rypma B, Gabrieli JDE, Tversky B, Glover GH (1999) Imagined transformations of bodies: an fMRI investigation. Neuropsychologia 37:1029–1040
5. Zacks JM, Ollinger JM, Sheridan MA, Tversky B (2002) A parametric study of mental transformations of bodies. NeuroImage 16:857–872

6. Zacks JM, Vettel JM, Michelon P (2003) Imagined viewer and object rotations dissociated with event-related fMRI. J Cogn Neurosci 15:1002–1018
7. Wraga MJ, Creem SH, Proffitt DR (2000) Updating scenes after object- and viewer-rotations. J Exp Psychol Learn Mem Cogn 26:151–168
8. Shepard RN, Metzler J (1971) Mental rotation of three-dimensional objects. Science 171:701–703
9. Parsons LM (1987) Imagined spatial transformation of one's body. J Exp Psychol Gen 116:172–191
10. Zacks JM, Tversky B (2005) Multiple systems for spatial imagery: transformations of objects and bodies. Spat Cogn Comput 5:271–306
11. Alivisatos B, Petrides M (1997) Functional activation of the human brain during mental rotation. Neuropsychologia 35:111–118
12. Barnes J, Howard RJ, Senior C, Brammer M, Bullmore ET, Simmons A et al (2000) Cortical activity during rotational and linear transformations. Neuropsychologia 38:1148–1156
13. Bonda E, Petrides M, Frey S, Evans A (1995) Neural correlates of mental transformations of the body-in-space. Proc Natl Acad Sci U S A 92:11180–11184
14. Carpenter PA, Just MA, Keller TA, Eddy W, Thulborn K (1999) Graded functional activation in the visuospatial system with the amount of task demand. J Cogn Neurosci 11:9–24
15. Cohen MS, Kosslyn SM, Breiter HC, DiGirolamo DJ, Thompson WL, Anderson AK et al (1996) Changes in cortical activity during mental rotation: a mapping study using functional MRI. Brain 119:89–100
16. Gauthier I, Hayward WG, Tarr MJ, Anderson AW, Skudlarski P, Gore JC (2002) BOLD Activity during mental rotation and viewpoint-dependent object recognition. Neuron 34:161–171
17. Harris IM, Egan GF, Sonkkila C, Tochon-Danguy HJ, Paxinos G, Watson JDG (2000) Selective right parietal lobe activation during mental rotation: a parametric PET study. Brain 123:65–73
18. Richter W, Somorjai R, Summers R, Jarmasz M, Menon RS, Gati JS et al (2000) Motor area activity during mental rotation studied by time-resolved single-trial fMRI. J Cogn Neurosci 12:310–320
19. Shelton AL, Pippitt HA (2006) Motion in the mind's eye: comparing mental and visual rotation. Cogn Affect Behav Neurosci 6:323–332
20. Tagaris GA (1998) Functional magnetic resonance imaging of mental rotation and memory scanning: a multidimensional scaling analysis of brain activation patterns. Brain Res Rev 26:106–112.
21. Vanrie J, Béatse E, Wagemans J, Sunaert S, Van Hecke P (2002) Mental rotation versus invariant features in object perception from different viewpoints: an fMRI study. Neuropsychologia 40:917–930
22. Zacks JM (2008) Neuroimaging studies of mental rotation: A meta-analysis and review. J Cogn Neurosci 20:1–19
23. Aguirre GK, D'Esposito M (1997) Environmental knowledge is subserved by separable dorsal/ventral neural areas. J Cogn Neurosci 17:2512–2518
24. Bonda E, Petrides M, Ostry D, Evans A (1996) Specific involvement of human parietal systems and the amygdala in the perception of biological motion. J Cogn Neurosci 16:3737–3744
25. Cohen JD, MacWhinney B, Flatt M, Provost J (1993) PsyScope: a new graphic interface environment for designing psychological experiments. Behav Res Methods, Instrum Comput 25
26. Shelton A, Lau Y, Zacks J, Yoon BC (2008) The opportunistic use of reference frames for rotating scene stimuli. J Vis 8:739

Author Index

J. S. Gero (ed.), *Studying Visual and Spatial Reasoning for Design Creativity*,
DOI 10.1007/978-94-017-9297-4

MIX
Papier aus verantwortungsvollen Quellen
Paper from responsible sources
FSC® C105338

If you have any concerns about our products,
you can contact us on
ProductSafety@springernature.com

In case Publisher is established outside the EU,
the EU authorized representative is:
Springer Nature Customer Service Center GmbH
Europaplatz 3, 69115 Heidelberg, Germany

Printed by Libri Plureos GmbH
in Hamburg, Germany